KU-075-390

PREFACE

The traditional subject of Dynamics of Machines has always suffered from the disadvantage of having no apparent underlying theme by which its various facets may be connected. The reason for writing this book has been to try to overcome this problem by suggesting a simple rationalisation scheme based on the output characteristics of power converters and the matching of them to their loads. These output characteristics are defined in terms of a convenient and well-known rate parameter (current I, flow-rate Q, speed N or heat transfer rate per unit temperature difference q/ψ) and associated parameters which, when multiplied by the appropriate rate parameters, give energy flow-rate or power. These associated parameters are also familiar and are, respectively, voltage differential V, pressure differential P, torque T and temperature differential ψ and may be termed potential parameters.

Transmission devices are introduced to match a converter to its load and storage elements to store energy of a kinetic (rate) form or a potential (latent) form for use at a later point in time. As a result, the need for a gearbox, a slipping element and an energy store are elucidated and the epicyclic gearbox, for example, is developed in response to a need for a particular kind of 3-terminal transmission element.

In this book the subject of Mechanical Systems is presented on the basis of component input-output relationships and as such it restricts itself in a concise way to the consideration of lumped-parameter problems and to the interrelationships between lumped components or 'black-boxes' in an engineering-system. Investigation into the 'black-boxes' is kept to a minimum and the principles involved are explained as far as possible from physical and qualitative reasoning. The presentation is design-oriented with an emphasis on matching the man-made power interfaces and delving into the 'black-boxes' is left to appropriate specialist works. For example consideration is not given here to the applied mechanics (including vibrations) describing the internal operation of a system element.

All power converters which feature mechanical, fluid or thermal power at their input or output are construed to be within the scope of this book, and consideration is given in an elementary way to electric motors and generators. Since a large part of the engineer's work is to make a sensible choice of converter, some mention of electro-mechanical converters was considered essential. It also allows reference to an increased number of analogies and brings with it greater flexibility in investigating design possibilities as it often happens that new designs are initiated to satisfy a certain analogy.

It is hoped that this book will induce in the reader the habit of paying attention to the forms of characteristic appropriate to the range of elements from which he may choose when designing a mechanical engineering system.

The Author would like to express his appreciation of the discussions held with his colleagues in the School of Engineering and Applied Sciences at the University of Sussex which were of great assistance in organising the subject matter and devising the many examples.

CHAPTER 1

THE ENGINEERING SYSTEM

Engineering is concerned with power systems. The particular processes which energy may undergo in these systems are those of conversion, transmission and storage. For effective operation of the system the energy must also be suitably controlled and contained. Let us consider each of these facets in turn.

Energy can exist in many forms but for the vast majority of engineering purposes these may be grouped under four main headings - Mechanical, Fluid, Thermal and Electrical. Conventional energy-conversion plants consist of a boiler raising steam (thermal to fluid conversion), a steam-turbine (fluid to mechanical conversion) and an alternator (mechanical to electrical conversion) and are mechanically contained. For high efficiency a high fluid temperature is required but the effect of thermal stresses in combination with the other stresses in the mechanical containment limit this efficiency to about 35%. Hence other forms of conversion are being studied, notably the thermionic effect, the thermo-electric effect and the magnetohydrodynamic (M.H.D.) process, although their prospects at present are uncertain.

Power may be transmitted in all the forms mentioned; by mechanisms and machines (mechanical), by fluid pressure as in hydrostatic drives (fluid), by heat transfer processes (thermal) and by transmission lines and radio waves (electrical). All the four forms are also evident in the storage of energy, from springs (mechanical), through accumulators (hydraulic and electrical) to heat storage elements (thermal). Control may be achieved by electrical circuits, by mechanical linkages, fluidic devices, or in a few cases, thermal expansion processes. The only area in which all four forms are not evident is that of containment which tends for obvious reasons to be almost entirely dominated by mechanical means. However, an electrical method has been tried in the form of the 'pinch-field' as used in the containment of ionised gas in fusion experiments and oil-lubricated bearings may be regarded as a form of liquid containment since they rely on hydrodynamic pressure generated in an oil film to support an external load. The ultimate containment is, however, invariably mechanical.

1.1 Power Conversion Parameters

Consider first power converters. Remembering that there are four convenient forms of energy for engineering purposes, these give rise to twelve classes of power conversion, examples being indicated in Fig. 1.1a.

It should be noted that in each of the power ellipses the appropriate parameters of the *output* characteristics are given. For example, if we consider the mechanical form of power, the output characteristic of a motor, turbine or thermal-expansion engine is conveniently depicted on a plot of torque (or moment) versus speed. Such a plot indicates the importance of output speed on the magnitude of the torque delivered. It should be noted that the product torque times speed = energy-rate or power. Similarly, in the other power ellipses, the products voltage times current, pressure times discharge and temperature difference times 'specific' heat-flow each equal power. In the last 'specific' heat-flow is defined simply as the quotient heat-flow per unit temperature difference.

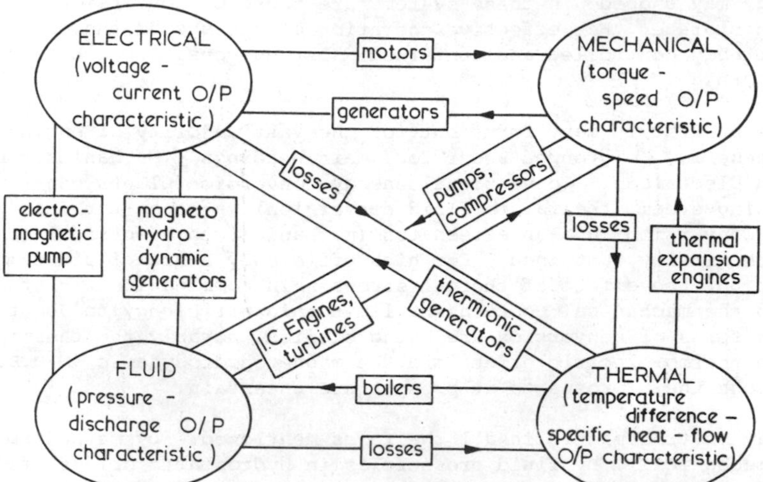

Fig. 1.1a. Conversion between various energy forms (based on a representation in 'Engineering Systems Analysis' by A. G. J. Macfarlane (Harrap, 1964))

The first parameter in each of these products may be regarded as a 'potential' parameter, since it has a latent power capability. The second variable is a rate parameter, by which the power ultimately manifests itself. To some extent this is an arbitrary categorisation, as it could be argued that voltage, for example, arises from a rate effect - that of cutting lines of magnetic flux. Also speed might be regarded as a potential parameter with reference to a stationary earth. For the purposes of engineering, however, the most consistent categorisation would appear to be in terms of potential and rate parameters *as they appear at the output* of a converter.

1.2 Introductory Example: The internal-combustion engine

As our first example let us consider the case of the automobile and its power converter (or engine) which converts the fluid power obtained from burning fuel to mechanical power at its output shaft (Fig. 1.2a). The fuel-inlet and exhaust valves are operated from the output (or crank) shaft in such a way that the valves open and close at the requisite points in the cycle to allow the induction of fuel-air mixture prior to combustion and the removal of the burnt products after combustion. Fig. 1.2b shows the resulting torque-speed curves for an automotive engine and it will be noted that the output torque is variable. Consider the full throttle condition, which gives maximum torque at any given speed. At low speeds the torque is less than its maximum value owing to inadequate mixing and owing to the valve timing which is usually designed for a moderately high speed; the torque reaches a maximum and afterwards decreases more and more rapidly as the speed increases. This decrease at high speeds is due to increasing friction losses and to less complete filling of the cylinders owing to the greater difficulty the pistons have in drawing the fuel-air mixture at high speed through the induction passages and valve ports (Fig. 1.2a).

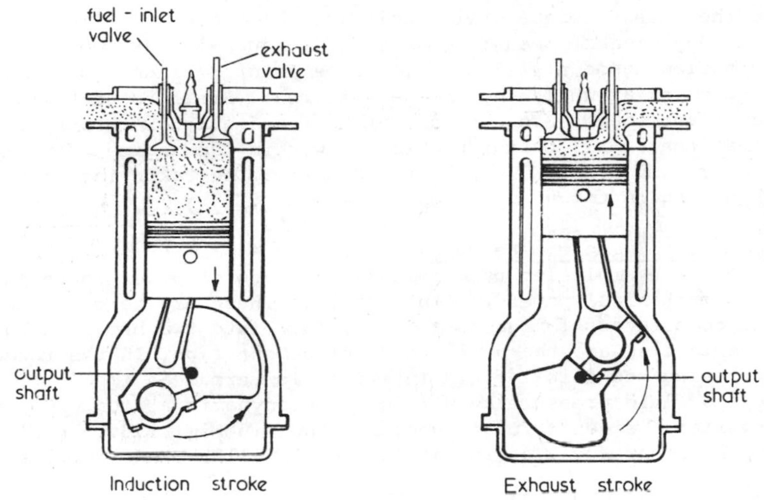

Fig. 1.2a. The automotive engine

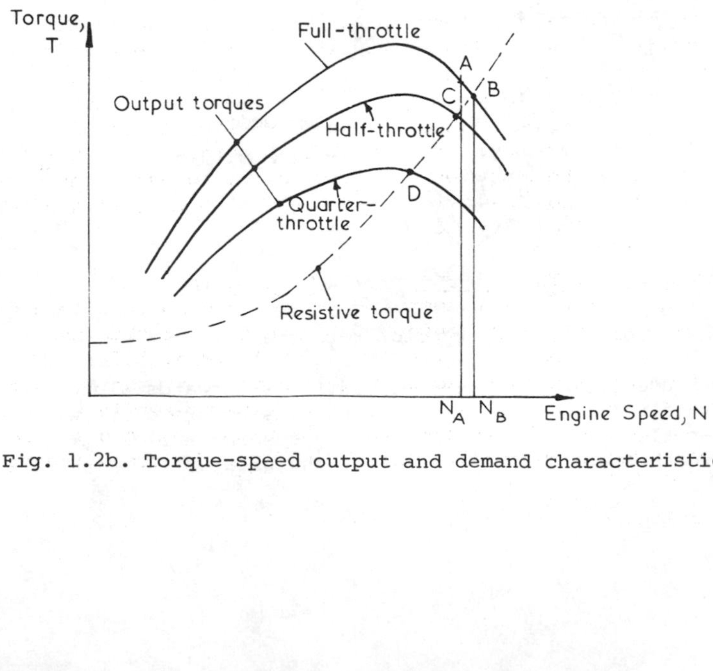

Fig. 1.2b. Torque-speed output and demand characteristics

Now assume that the automobile is travelling along a road. The faster it travels the more resistance it encounters owing to the wind and the road surface. The power required to overcome the resistive frictional demands is converted to heat by shearing action with the air and ground and is then transmitted to the atmosphere owing to the resulting temperature difference. There will be a relation between the car's forward speed and the resistance, which may be represented as a relation between engine speed and resistive torque felt by the engine. As such the relation may be plotted on the axes of torque and speed and will take the form shown dashed in Fig. 1.2b, in which the intercept on the torque axis accounts for any incline which the car may be climbing. Assume that the engine is running at full-throttle at point A on its characteristic. For the engine speed N_A the engine torque thus exceeds the resistive torque and the excess torque will accelerate the engine until a speed N_B is reached. At any higher speed the resistive torque will exceed the engine torque and a slowing-down will ensue. Hence point B is an equilibrium point or operating point at which the engine will run. Similar operating points at lower torques and speeds are obtained at half and quarter throttle. It should be noted that there will be torque demands within the engine, for example oil pump and water pump demands, which are assumed to be taken into account in constructing the engine torque-speed characteristic.

1.3 Introductory Example: The pump

As our second example let us consider the class of power converters which convert from mechanical power to fluid power. These are termed pumps for liquids and compressors for gases and they fall into two broad categories - the rotodynamic type and the positive displacement type. In the rotodynamic type the rotating mechanical-fluid interface is termed an impeller. Its purpose is to generate fluid pressure by virtue of rotodynamic action. One example of such a converter, a centrifugal pump, is shown in Fig. 1.3a. Fluid pressure is built up by centrifugal action along each impeller blade increasing from

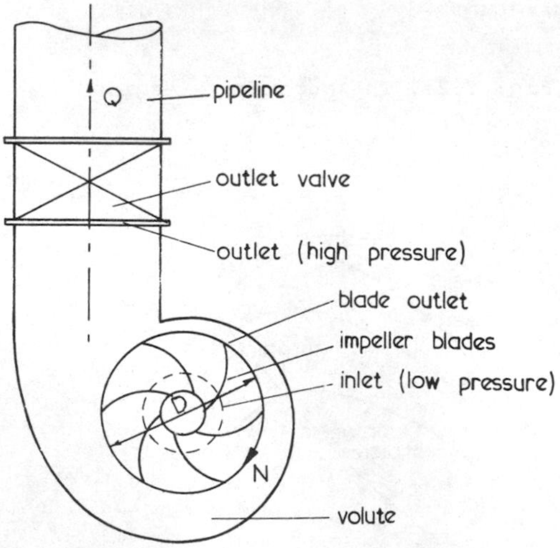

Fig. 1.3a. One type of rotodynamic pump

blade-inlet to blade-outlet. Thus no fluid flows outwards along the blade when the outlet-valve is closed and hence the casing or volute will not burst. When the blade-outlet pressure is relieved by opening the outlet valve a flow takes place through the pump and the valve; the greater the kinetic energy of flow the less the blade outlet pressure energy. The purpose of the volute or diffuser is to recover further pressure energy from any excess velocity energy in the fluid at the outlet tips of the impeller blades.

In accordance with the above reasoning we may write,

$$\text{outlet pressure } P = k_1 - k_2 Q^2,$$

where k_1 is the centrifugal pressure at closed valve and $k_2 Q^2$ is the pressure loss due to the output flow, Q. Since k_1 represents a centrifugal effect it is proportional to the square of the speed N of the impeller and we may also write,

$$P = k_3 N^2 - k_2 Q^2 \tag{1.3.1}$$

The pressure-discharge (P-Q) characteristic is thus of the falling variety shown by curve (i) in Fig. 1.3b. Rotodynamic pumps are widely used for

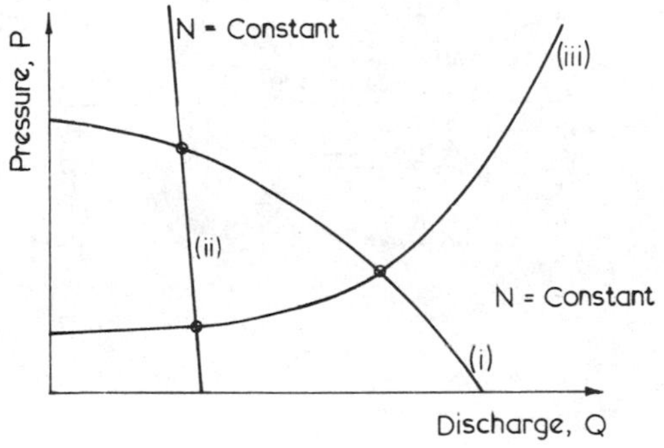

Fig. 1.3b. Pressure-flow characteristics

handling low-viscosity fluids and 95% of industrial fluids are transported by this means. Deliveries can range from a few litres/s to 2000 or more. Rotodynamic compressors are used for compressing air and other gases.

Positive displacement pumps fall into the broad categories of reciprocating pumps and rotary pumps and Fig. 1.3c shows diagrammatically a rotary type of mechanical-fluid converter. Here discrete pockets of liquid (shown shaded), effectively isolated from each other, are passed along from right to left. Thus each pocket is positively displaced forward irrespective of any pressure difference relative to the previous pocket. Thus, if the outlet valve were closed the result would be a theoretically infinite pressure and the casing may well burst. In practice some reverse leakage does take place between pockets but not sufficiently to allow the outlet valve to be closed with safety. Overload protection in the form of a pressure relief valve (Fig. 1.3c) is usually employed in case of inadvertent closure of the outlet valve.

Assuming no leakage and hence no losses, we may thus write,

$$Q_1 = k_5 N, \tag{1.3.2}$$

where Q_1 is the output flow, assumed incompressible. Allowing for leakage, we may write leakage flow Q_L, dependent upon developed pressure P, as

$$Q_L = k_4 P$$

and hence net flow out of the pump,

$$Q = Q_1 - Q_L \tag{1.3.3}$$

or

$$Q = k_5 N - k_4 P \tag{1.3.4}$$

This gives a constant-speed characteristic of the form of curve (ii) shown in Fig. 1.3b. Reciprocating pumps have a similar characteristic by virtue of the fact that they dispatch discrete pockets of liquid into a pipeline which would burst if the outlet valve were closed.

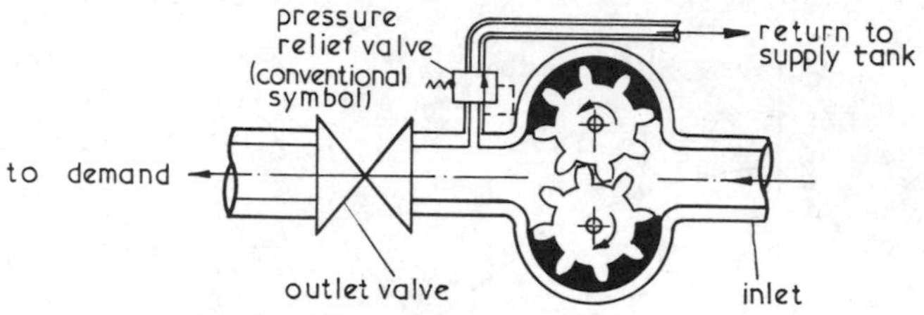

Fig. 1.3c. The positive displacement pump

When the pumped fluid is air or gas then positive displacement compressors are used. Here a static pressure rise is obtained in a system by allowing successive volumes of air to be aspired into and exhausted out of a closed space into the system by means of the displacement of a moving member. This member either reciprocates as in piston and diaphragm compressors, or rotates as in screw compressors and Roots blowers.

The pipeline into which the pump or compressor is feeding together with the outlet valve will have a resistance characteristic depending upon the demand. Such a characteristic is shown as load curve (iii) on Fig. 1.3b and in general will intersect the vertical axis with an intercept corresponding to the height through which the fluid is to be pumped. Friction in the pipeline increases with increasing flow rate and requires an increasing pressure to overcome it. This accounts for the curved portion of the characteristic. The operating point is determined by the intersection of the appropriate pump or compressor output characteristic (i) or (ii) and the load characteristic (iii). If the pressure corresponding to the operating point is higher than the setting of any pressure-relief valve then such an operating pressure will of course never be achieved.

1.4 Introductory Example: The electrical generator

Let us now consider a converter employing mechanical energy to produce electrical energy. The simplest of this class of converter is the direct-current (d.c.) generator, an exploded view of which is shown in Fig. 1.4a. It consists of a frame and a rotor or armature, the former containing windings or

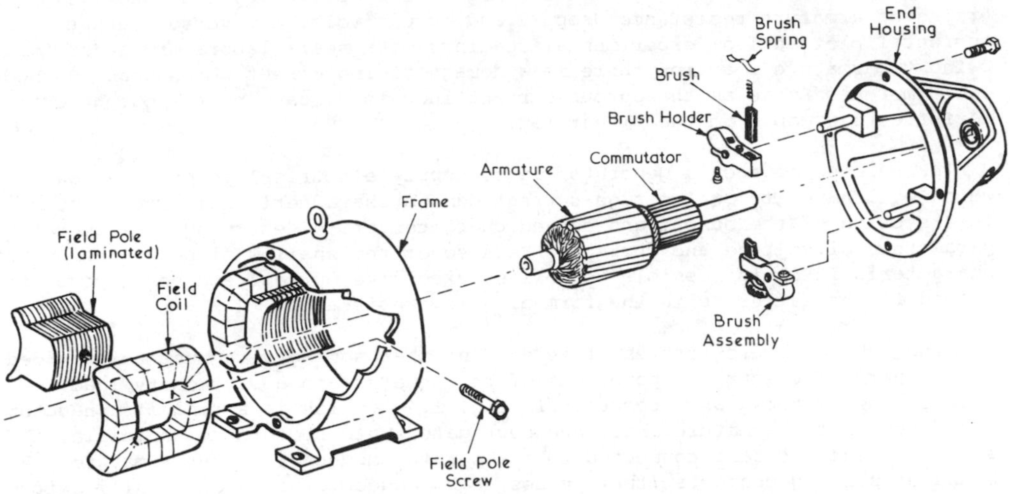

Fig. 1.4a. The d.c. generator

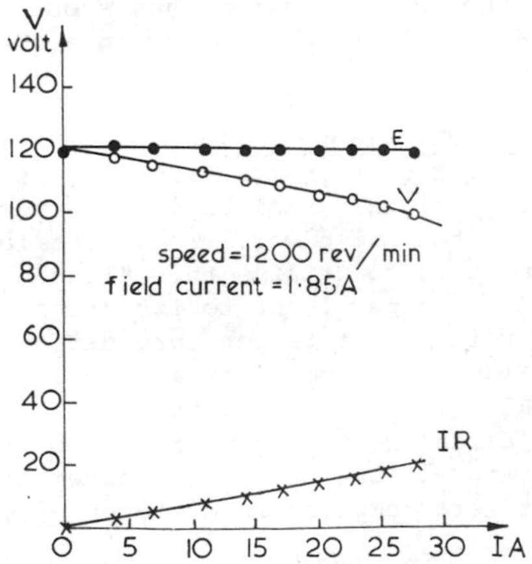

Fig. 1.4b. D.c. generator characteristics

coils which provide a magnetic field within which the armature is rotated. An electro-motive force or e.m.f. E is thereby generated in the conductors of the armature. The available voltage is less than E due to the small resistance drop in the armature windings and mathematically, we may write for any armature

$$V = E - IR \qquad (1.4.1)$$

where E is the generated e.m.f. and is proportional to input speed N and to the field flux; R is the small resistance of the armature windings. For a constant field current I_f, supplied from an independent source, the voltage-current output characteristic of this d.c. generator for a given rotor speed is thus of a falling variety with low negative slope. Fig. 1.4b shows typical graphs of armature resistance drop IR and output voltage V versus output current I plotted from experimental readings on a small laboratory generator. Owing to armature reaction there is a demagnetising effect which weakens the field more and more as the output current increases, causing the output characteristic to droop from the linear form.

The purpose of such a generator is to supply electrical power to a load which will have its own voltage-current demand characteristic. The point of intersection of the output and demand characteristics determines the operating parameters of voltage and current. Because of the shallow slope of the output characteristic it is possible to deliver excessive current and hence power to a load if the latter is in the form of a low resistance.

Since the armature rotates it is evident that some device must be provided to deliver the electrical power out of the armature to a stationary load. This device is known as a commutator (Fig. 1.4a) which is a ring of conductors connected to the armature shaft and segregated into sectors, each insulated from the next and each connected to a separate conductor in the armature. By means of sliding contacts (the brushes) each conductor is sequentially switched to the d.c. load so that, in its passage through the field, each provides a current and voltage in the same direction. A fair analogy to the commutator of a d.c. generator is the valve-gear of a steam engine. The main purpose of the valve-gear is to allow steam to be admitted alternately to each side of the piston as it drives the output shaft. Thus sequential steam forces in <u>different</u> directions are made to produce power in a requisite direction.

1.5 Introductory Example: The electric fire

The VI demand characteristic of an electric fire is determined by its electrical resistance. Because of this resistance a heat-loss ensues in a conversion from electrical to thermal power. The electric fire thus acts as a converter whose requirement is to utilise this "loss" to full effect in, say, heating a room. As such it is pertinent to investigate its output characteristic described in coordinates of temperature difference ψ between the room temperature and the outside air temperature and heat dissipation q in the form of radiant, convective and conductive heat transfer. Just as electrical voltage is measured above 'earth' so that temperature of the room is here measured above the outside temperature. The output characteristic is of the form shown in Fig. 1.5a. Thus, for zero temperature difference ψ, the heat transfer q will be a maximum.

Let us now inspect the demand characteristic imposed by the room on the electric fire. An increase in room temperature implies an increase in ψ together with an increase in heat transfer q due to the increased heat losses from the walls of the room. Hence the demand characteristic is of a rising form. Its intersection with the output characteristic of the electric fire gives the operating temperature difference and heat transfer from the fire.

To be consistent with the potential and rate parameters used in the other introductory examples we could equally have plotted heat flow per unit temperature difference between the room and the outside air versus this same temperature difference. Their product would then have been equal to power.

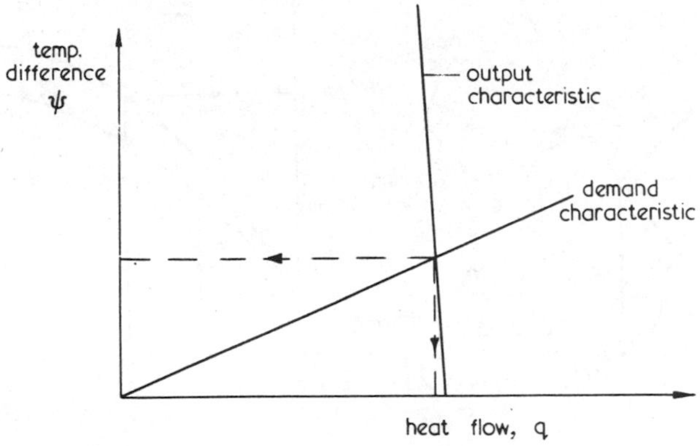

Fig. 1.5a. Output and load characteristics of an electric fire

1.6 Introductory Example: Power converters in series

A great deal of the work of an engineer is devoted to designing a system whose purpose is to supply power to fulfil a given task. For example, it may be necessary to pump a fluid from one point to another at a fixed rate. If the fluid is contained in a pipe then a certain amount of power is needed to overcome the frictional resistance while the fluid is in motion in the pipe. The engineer's job is then to decide on the type of power converter which he must employ and whether transmission and/or storage devices are also needed. Consider for example a pipeline whose purpose is to convey water from one point to another at the same level. It will probably be convenient to use an electric motor as the power supply. At the actual pumping point, that is at the mechanical-fluid interface a mechanical-fluid converter would be used in the form of a pump impeller. Remembering that the *original* energy source may be coal in a conventional power station, the appropriate power conversions are, referring to Fig. 1.1a, THERMAL - boiler - FLUID - steam turbine - MECHANICAL - generator - ELECTRICAL - motor - MECHANICAL - pump - FLUID - losses - THERMAL.

Two important points should be noted,

(1) The initial and final energy states are thermal, which is true of all complete cycles, which do not terminate in a potential energy store.

(2) The intermediate electrical power state is required to effect efficient transmission through the grid system.

The various converter output characteristics are presented in the form of graphs in Fig. 1.6. The method adopted to deduce the pump delivery pressure and motor torque and speed as well as the required supply voltage and current is completely general, and no further attention will be paid at the moment to the physical reasons which decide the shapes of the characteristics. It is however necessary to define a term which will be used frequently in this book, that of efficiency. This is the ratio ouput power/input power and is usually plotted against the output rate parameter to give an efficiency characteristic.

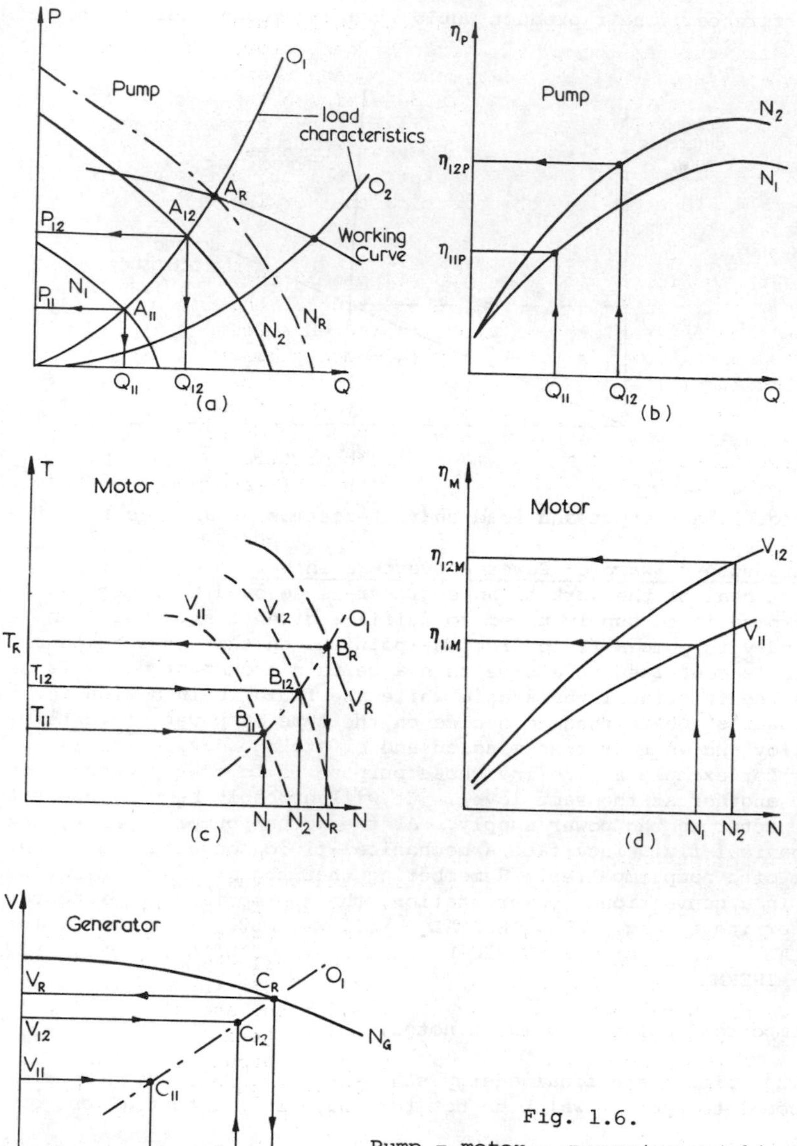

Fig. 1.6.
Pump - motor - generator matching

Consider Figs. 1.6a and 1.6b in which we are presented with pressure P - discharge Q characteristics and with efficiency η_p - discharge Q characteristics for a centrifugal pump having different drive-motor speeds N. Let us assume that the pump is being used to circulate fluid in a pipeline whose resistance must be overcome by the pump delivery pressure P which in turn determines the volume flow Q. The product PQ affects the pump motor's speed N, output torque T and power requirements which in turn create a demand on the voltage V and current I supplied from the generator. The magnitude of this demand depends upon the pump efficiency η_p and upon the motor efficiency η_M. Thus the outlet valve opening O , in determining the load characteristic (Fig. 1.6a), ultimately makes itself felt on the generator supply. Extrapolating further back we could even argue that it makes itself felt on the rate of supply of coal needed in the power-station boiler, although this effect is infinitesimal. How do we then determine the working values of the parameters P, Q, N, T, V and I?

Returning to Fig. 1.6a we can superimpose on the P-Q characteristics of the pump-supply corresponding P-Q characteristics of the pipeline demand. These will depend on the valve opening O. Characteristics corresponding to two different openings O_1 and O_2 are given. Such characteristics would be obtained from a knowledge of the valve and pipe dimensions and friction data. We require to find the delivery pressure and volume flow which prevail under the imposed condition corresponding to valve opening O_1 say. Consider the point of intersection A_{11} of the O_1 demand characteristic and the pressure-flow supply characteristic for speed N_1. Let the corresponding delivery pressure and volume flow be P_{11} and Q_{11} and the efficiency of power conversion from impeller input to output be η_{11P} (Fig. 1.6b). Then for the pump,

$$\eta_{11P} = P_{11}Q_{11}/T_{11}N_1,$$

or

$$T_{11} = P_{11}Q_{11}/N_1\eta_{11P},$$

where T_{11} and N_1 are the output torque and speed respectively of the pump-motor. Similarly, for point of intersection A_{12},

$$T_{12} = P_{12}Q_{12}/N_2\eta_{12P}.$$

In general the pump efficiencies η_{11P} and η_{12P} are obtained from independent tests and are usually given in characteristic curves such as Fig. 1.6b. Turning now to Fig. 1.6c, motor torques, T_{11} and T_{12} may be inserted on the T axis bounding the motor characteristics, and, together with the already prescribed motor speeds N_1 and N_2, give points B_{11} and B_{12}. Similar points may be found by employing different speeds in Figs. 1.6a, b and c. A curve may now be drawn through the points B_{11}, B_{12}, etc. and given the symbol O_1 (Fig. 1.6c) since it exists for constant valve opening O_1. Consider now the operation of the pump motor. The torque-speed output characteristic of the motor (Fig. 1.6c) will depend on the supply voltage V from the generator via the mains, and each point B_{11} etc. will determine a corresponding voltage V_{11} etc. In Fig. 1.6c two such voltage curves V_{11} and V_{12} are shown. The question now arises as to how we determine the precise voltage-current point at which the motor will operate for the valve opening O_1. To do this we must proceed one step further and consider the generator which generates the mains current I and voltage V in the power station. Its action is similar to that of the centrifugal pump in that it "discharges" a current I at a mains "delivery pressure" V depending on the generator speed N_G. For a constant value of N_G we might have a voltage-current characteristic as shown in Fig. 1.6e. In British power stations N_G = 3000 rev/min, while in the USA the figure is 3600 rev/min. Let us now superimpose on Fig. 1.6e the information we have already gained from Figs. 1.6a, b and c. Since both the products TN and VI denote power we may say that, for the motor characterised by Figs. 1.6c and d,

$$I_{11} = T_{11}N_1/V_{11}\eta_{11M}$$

$$I_{12} = T_{12}N_2/V_{12}\eta_{12M}, \text{ etc.}$$

In these equations η_{11M} etc. are the motor efficiencies which again can be independently ascertained. Thus points C_{11}, C_{12}, etc. can be located in Fig. 1.6e and joined to give curve O_1. This represents the "demand" curve required on the generator. The intersection of the "supply" curve and this demand curve gives the operating point C_R, since at such a point supply and demand must be equal. The point C_R defines the required current I_R and voltage V_R demanded of the generator. The curve corresponding to V_R can now be used in Fig. 1.6c in conjunction with the already existing O_1 curve to give the operating point B_R for the motor. This point is associated with a speed N_R which in turn is located as a curve on the pump characteristics of Fig. 1.6a, and where this curve intersects the O_1 curve gives the operating point A_R for the pump. The whole process can of course be repeated for other valve openings such as O_2, etc. and a working curve constructed.

In many pump applications a constant supply voltage may be assumed and the torque-speed output curves for the motor are often such as the give very little speed variation for a large variation in supply torque. In other words the motor may be regarded as a constant-speed device. This means that the working curve of Fig. 1.6a is virtually identical with an appropriate constant-speed curve and interest is restricted to Fig. 1.6a. There are, however, many alternative forms of drive for a rotodynamic pump such as reciprocating engines and steam turbines, and here interest would have to be focused at least on figures equivalent to Figs. 1.6a, b and c.

Incidentally, it should be noted that efficiency characteristics such as in Fig. 1.6b can be accommodated in output characteristics such as in Fig. 1.6a by the back-plotting of iso-efficiency curves. For example, if we select an efficiency η_{11P}, then a table of values of N and Q can be constructed, from which corresponding values of P can be deduced (Fig. 1.6a). A succession of points is thus defined in axes of P and Q, forming an iso-efficiency curve.

CHAPTER 2

POWER CONVERSION

2.1 General Forms of Converter Output Characteristic

In the introductory examples of chapter 1 we met several output characteristics which may be conveniently categorised into four basic forms. The most important features of a characteristic are the value of the 'potential' parameter (torque, pressure, voltage or temperature difference) when the 'rate' parameter (speed, flow, current or specific heat-flow) is zero and the value of the slope over the major part of the operating range. Thus for the automotive engine we saw the form (i) (Fig. 2.1a) in which T is zero when N is zero and whose slope ranges from positive values to low negative values. By contrast, in the rotodynamic pump we encountered the form (ii) which illustrates a positive value of potential parameter at zero rate parameter and also intersects the rate parameter axis. In the positive-displacement pump and the electric-fire we found form (iii) exhibiting a well-defined intercept on the rate parameter axis and having a very large negative slope, and in the d.c. generator we saw form (iv) having a well-defined intercept on the potential parameter axis and a low negative slope. Form (i) occurs with converters which provide their own operating environment; they might be termed self-induction machines. Forms (ii), (iii) and (iv) can only arise for converters which are not self-inducing.

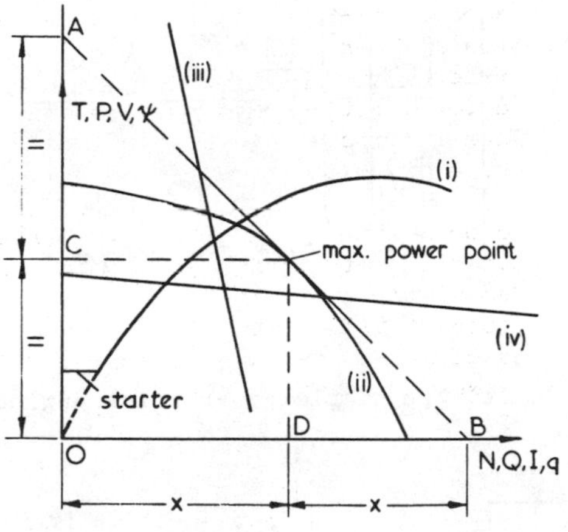

Fig. 2.1a. General forms of output characteristic

Type (i) characteristic. By way of elucidating the above definitions consider again the automotive engine. The torque generated by one combustion stroke is used to draw in and compress the charge of fuel-air mixture for the next combustion stroke. The engine may be represented diagrammatically as shown in Fig. 2.1b. Because of this action the engine can generate no torque at zero speed; in other words it has no 'starting-torque'. The same feature is present in the simple gas turbine engine. Similarly, for the series-excited generator there is no 'starting-voltage'.

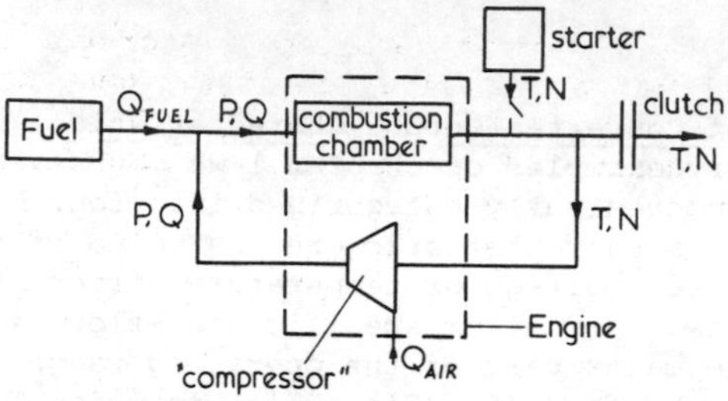

Fig. 2.1b. The automotive engine

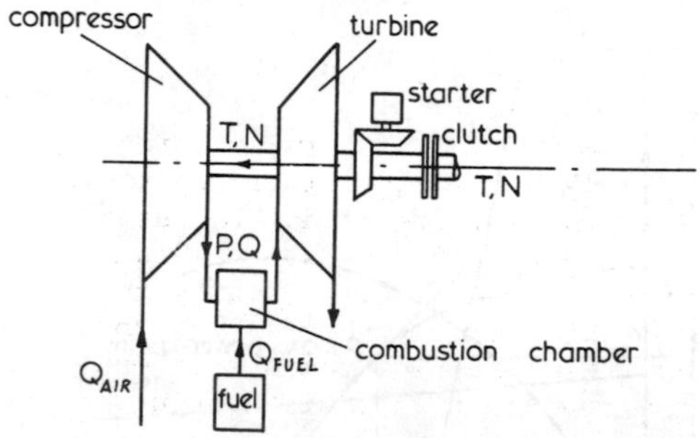

Fig. 2.1c. The simple gas-turbine engine

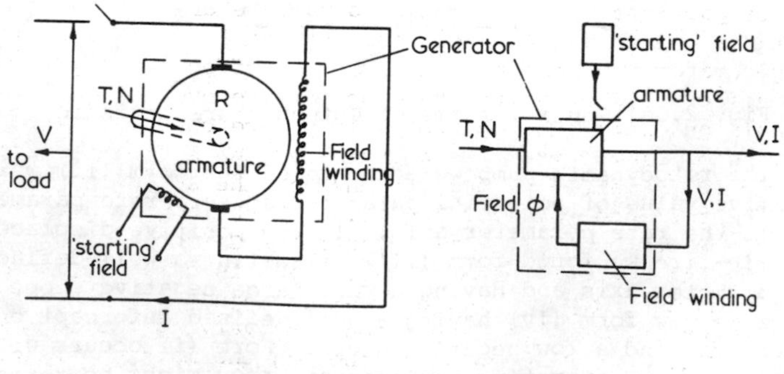

Fig. 2.1d. The series generator

The need for the intermittent use of the piston as a compressor in the reciprocating engine is obviated in the simple gas turbine by the provision of a permanent compressor. This is a series of bladed discs which raise the air pressure by rotodynamic action. The complete machine may be represented diagrammatically as shown in Fig. 2.1c. The rôle of the combustion chamber is to add heat energy to the fluid so that it expands as it impinges onto the blades of the turbine. This elongates the pressure-flow output characteristic of the compressor in the Q direction and hence gives extra power to the turbine. A reduction in load speed implies a reduction in compressor speed, thus reducing the flow of air to the combustion chamber, and this is accompanied by a fall in output torque.

Similarly in the direct-current series-generator the winding providing the magnetic field necessary for the operation of the generator is in series with the armature and utilises the output current I. The d.c. series-generator may be represented as shown in Fig. 2.1d and reference to equation (1.4.1) indicates that the following equations hold,

$$V = E - IR$$

$$E = kNI, \qquad (2.1.1)$$

since I serves as the field current. Hence

$$V = kNI - IR$$

$$= I(kN - R), \qquad (2.1.2)$$

where R consists of the field resistance in addition to the armature resistance. Equation (2.1.2) gives a characteristic of the same rising form as that of the automotive engine. In fact, due to saturation of the field there is a departure from the linear form expressed by equation (2.1.2) and the voltage-current characteristic bends over much as the torque-speed characteristic of the automotive engine bends over (Fig. 1.2b). From equation (2.1.2) it can also be seen that the effect of speed N is much the same as the effect of throttle opening on the automotive engine output characteristic.

There are several more favourable alternatives than the series-generator for direct-current power generation, separate and parallel (or shunt) field excitation giving a positive output voltage at zero output current. The d.c. series-generator is used for boosting purposes, that is to inject a voltage into a cable proportional to the preceding voltage drop in the cable caused by its resistance. Such a generator thus appears as a negative load to the main supply generator and may be regarded as an effective negative resistance. This has the effect of decreasing the slope of the overall load characteristic seen by the main generator and thereby allows more current to serve the load. Alternatively, when viewed from the actual load the effective generated output characteristic is raised by the presence of the series generator. To elucidate its performance consider Fig. 2.1e which shows a diagram of a laboratory test rig which embodies a main generator G_1 whose output to a load is to be boosted by the series-generator, G_2. Each is driven by a motor running at a set speed and the output voltage of G_1 is adjusted to 120V by setting its field current which is then left unchanged. By varying the load resistance, R the graphs of Fig. 2.1e were obtained showing the main generator output voltage V_1, the series generator output voltage V_2 and the boosted load voltage V_3 versus load current I. These illustrate the boosting effect of the series-generator in lifting the effective output characteristic from V_1 to V_3. The difference between the generated e.m.f. E_2 and output voltage V_2 is indicative of the internal voltage drop in the series-generator.

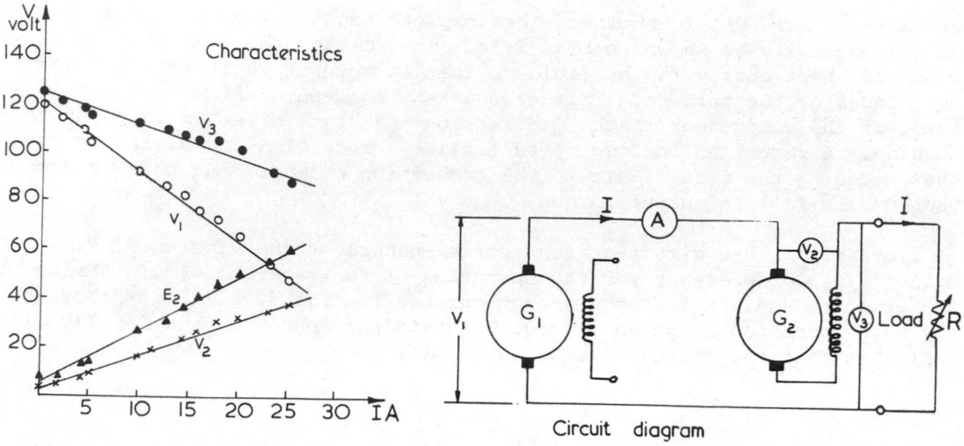

Fig. 2.1e. The series generator as a booster

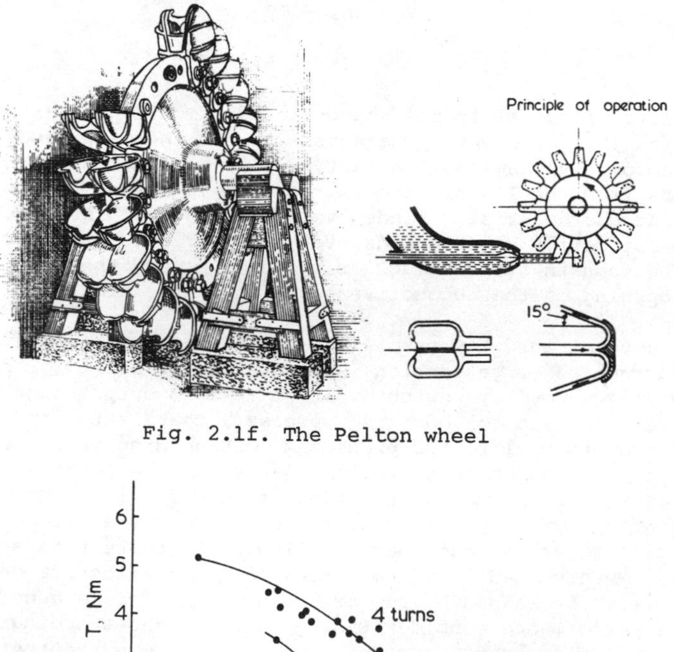

Fig. 2.1f. The Pelton wheel

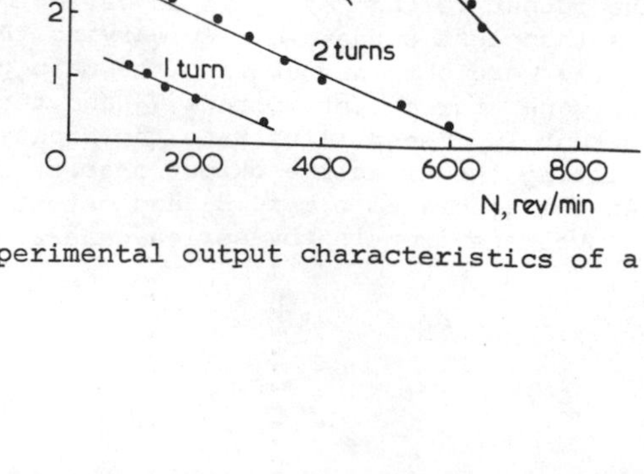

Fig. 2.1g. Experimental output characteristics of a Pelton wheel

In the case of the automotive and gas turbine engines the desirability of using energy in its most compact form, namely as fossil fuel, makes self-induction a paramount requirement. However, any self-inducing converter, by exhibiting no starting-torque, requires a separate means of starting. Explosive cartridges or starter motors connected to the output shaft are used for the piston engine and the simple gas turbine and temporary separate excitation of the field for the series-generator. The effect of a separate starting device on a torque-speed output characteristic of type (i) is shown in Fig. 2.1a.

Type (ii) characteristic. In the case of converter output characteristics having the form (ii) in Fig. 2.1a, the existence of a negative slope and the reduction of the potential parameter from some intercept value indicates that some 'slip' is taking place between the input and output rate parameters. For example, in the rotodyamic pump the input rate parameter is the speed N of the impeller, which is not completely reflected in the flow of fluid Q in the volute and slip is said to exist between the two. Broadly speaking, maximum slip (i.e. zero discharge) is a condition of maximum output pressure. Alternatively we may regard the condition of zero discharge as a condition of zero kinetic-energy (or fluid rate energy) and hence maximum potential energy.

Let us consider now other converters having a characteristic of the same form. Examples of such converters are the Pelton wheel, the water and steam turbines and the induction motor. In the Pelton wheel and in the water and steam turbines a moving fluid supplies energy to a mechanical rotor and generally speaking the greater the difference in speed or the 'slip' between the fluid and the rotor the greater the torque transmitted to the rotor. In the induction motor a moving electrical field supplies energy to a mechanical rotor and again, under normal running conditions, the greater the slip the greater the torque transmitted.

Consider first the Pelton wheel. This is a type of impulse turbine and is used where relatively high heads of water of 130 m or more are available. It utilises the impulsive force imposed by a jet of water on its blades or buckets to create a torque, these buckets being set around the rim of a rotor called a runner. Fig. 2.1f shows a Pelton wheel in the course of construction together with its principle of operation. The water usually leaves the bucket after it has been turned through about 165° by the bucket surface. The impulsive force and hence the torque decrease as the bucket speed approaches the speed of the water jet, giving a torque-speed characteristic which drops with increase in speed from an initial intercept on the torque axis. Fig. 2.1g shows experimental characteristics obtained from a small laboratory Pelton wheel and illustrates the basic form. The number of turns of the inlet valve is a measure of the jet velocity and hence of flow delivery rate. The Pelton wheel is one of many fluid-mechanical energy converters. The simple water-wheel (for heads of 6 to 10 m) falls into this category as does the Francis reaction turbine (for heads of about 45 to 80 m). The latter operates in the reverse way to a centrifugal pump. For low heads with very large flow-rates an axial-flow turbine is employed, as for example in tidal systems. All these turbines exhibit the same basic type of torque-speed characteristic.

Let us now consider the induction motor in which a rotor is supplied with power from a rotating electrical field. When a conductor is placed in the rotating field of a cylindrical stator fed with an alternating supply current, it is cut by the field and an alternating electro-motive force (e.m.f.) is generated in the conductor having the same frequency as that of the supply. If instead of the conductor a short-circuited turn spanning the stator diameter is placed in the field the e.m.f. generated in the conductors forming the sides of the turn causes a current to flow which is proportional to the velocity of the field relative to the conductor and each conductor experiences a force proportional to this current in the direction of the field. In its simplest form the induction motor rotor is of laminated iron and has slots containing copper bars, the ends of which are short-circuited by heavy copper; this is known as a squirrel cage rotor. As the speed increases, the relative motion (the 'slip') decreases and hence the e.m.f., current and torque decrease. The application of a load causes the speed to decrease and thus produces an increase in rotor e.m.f. and current; the speed then falls to a value at which the rotor current is sufficient to give the increased torque required. A typical output characteristic is plotted in Fig. 2.1h. The torque increases to a

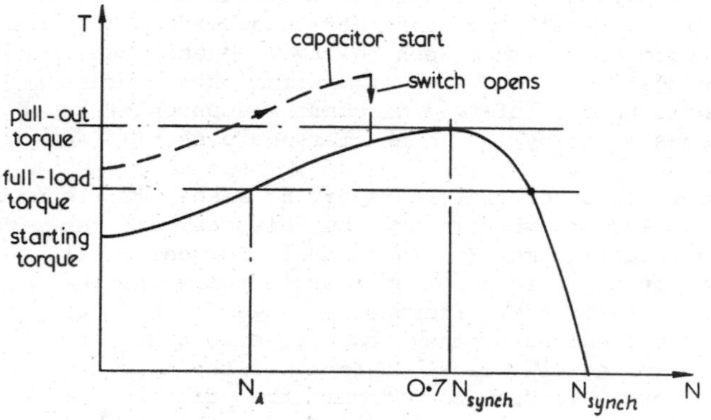

Fig. 2.1h. Induction motor output characteristic

maximum when the rotor has attained (in this case) about 70% of synchronous speed, which is determined by the frequency of the electrical supply. The torque then drops sharply to zero at synchronous speed which corresponds to zero slip. Normal load conditions are confined to the extreme right-hand portion which may be regarded as linear for most purposes and the full-load torque usually occurs with a slip of 3-5%, depending on the size of the machine. If a load torque is applied which is independent of speed and is gradually increased, the speed falls and the motor output torque increases until the maximum value, known as the pull-out torque, is reached at about 70% of synchronous speed; any further increase in load will then cause the motor to stop. If the demand torque were constant with speed at the full-load value, then the motor would not start, and hence some means must be provided to ensure starting. One method is to employ a capacitor at start up which later cuts out. Its effect is to give a curve of the form shown in Fig. 2.1h. It is interesting to note that, with the provision of an alternating current supply rather than a direct current supply, no commutator is now needed. In the same way no valve gear is needed in the turbine due to the provision of a mechanical rotor instead of the reciprocating piston of the automotive engine.

Two interesting variants of the induction motor are worth mentioning. These are the variable-frequency induction motor and the synchronous motor. The former acts exactly like the ordinary induction motor except that an increased speed range is afforded by varying the frequency of the supply to the stator by means of a frequency-changer. On the other hand the synchronous motor is used where an output speed exactly equal to (or synchronous with) the frequency of the a.c. supply to the stator is required. Such a machine is first run up to near synchronous speed under no load as an induction motor. The output shaft is provided with two slip-rings which are continuous rings around the shaft, in sliding contact, connected to each side of a d.c. supply. This d.c. supply supplies a magnetising current to the rotor, which takes the place of the current produced by slip in the ordinary induction motor, and a torque is produced by the action of the rotating stator field without the need for slip. Hence the rotor runs at synchronous speed, and its torque-speed characteristic is a vertical line perpendicular to the speed axis. If the machine is overloaded, then any significant fall in speed allows slip action to become predominant and the motor runs down as an induction motor. In small synchronous motors a permanent-magnet rotor is used instead of supplying a magnetising current through slip-rings. The synchronous motor has its counterpart in the alternator which converts mechanical power to alternating electrical power. This power is supplied through slip rings to load which requires it.

Type (iii) characteristic. Now let us consider converters having a characteristic of the form (iii) of Fig. 2.1a. These have been differentiated from form (ii) since they have a very high negative slope, and as a result any system of which these converters are a part is likely to need protection in the event of an output potential parameter such as pressure, torque or temperature becoming too great for the system to contain. For example we saw that the positive-displacement pump required a pressure-relief valve.

The d.c. separately excited motor also has such a falling characteristic. Its construction is exactly the same as that of the generator shown in Fig. 1.4a. However, instead of rotating the armature from an external source, that is supplying the armature with mechanical power, the armature conductors are now supplied with electrical power. When supplied with a field current the field coils and poles provide a magnetic field as for the case of the generator and the forces set up due to the current in the armature conductors cause the armature to rotate. The commutator serves the same purpose as in the generator and transfers each armature conductor to the external supply circuit as the armature rotates. As a result the force on the armature conductors produces a torque always in the same direction. However as soon as the armature rotates an e.m.f. is generated *due* to their motion in the field, just as in a generator. This reduces the effect of the current and is therefore known as a back-e.m.f. In order to maintain this current I the supply voltage V has to overcome this back-e.m.f. E as well as the small resistance R in the armature conductors.

Hence
$$V = IR + E. \tag{2.1.3}$$

As for the generator, E is proportional to armature speed N and to field current I_f, whilst the torque T produced at the armature is proportional to the armature current I as well as to I_f. Hence we may rewrite equation (2.1.3) as follows,

$$V = IR + k_7 I_f N,$$

whilst
$$T = k_7 I_f I.$$

Hence
$$V = T R/k_7 I_f + k_7 I_f N$$

or
$$T = V.k_7 I_f/R - N (k_7 I_f)^2/R. \tag{2.1.4}$$

Owing to the smallness of R equation (2.1.4) gives a steeply falling torque speed characteristic with a large intercept on the torque axis. This makes it similar to the pressure-flow characteristic of the positive-displacement pump. Consequently these motors may be regarded as constant-speed machines. When the resisting torque is increased during normal running the armature slows down just enough to decrease the back-e.m.f. and allow the increase in current required to produce the necessary increase in torque. If the load is decreased the existing torque accelerates the armature and the back-e.m.f. increases and reduces the current to that required by the new torque. Hence the motor, when connected to a constant voltage supply, takes the current required by the load. Thus, if it is overloaded, overheating of the machine windings may result. Hence the need for protection, which could take one of two forms - a fuse or overload release in the *supply* to the motor or a shear-pin in the *output* from the motor. The latter is analagous to the pressure relief valve at the output of the positive-displacement pump.

Type (iv) characteristic. The d.c. separately-excited generator (Fig. 1.4a) has a characteristic exhibiting a positive intercept on the potential parameter axis and a low negative slope. Just as the converters of type (iii) can be overloaded by an excessive potential output parameter so converters of type (iv) can be overloaded by an excessive rate output parameter such as current (causing overheating) or speed (causing high mechanical stresses in rotating components). The fuse or overload release on the *input* to the motor discussed above may be more logically regarded as protecting the *output* of the generator supplying the motor with electrical power. The output characteristics of d.c. generators may be varied by adopting a combination of fields in series or in parallel with the armature circuit. This is known as compound excitation and by this means varying values of slope may be obtained in the output characteristic. Some typical characteristics are shown in Fig. 2.1i.

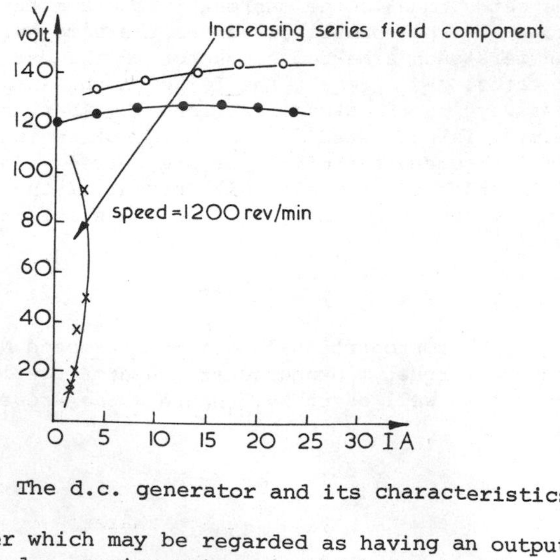

Fig. 2.1i. The d.c. generator and its characteristics

Another converter which may be regarded as having an output characteristic of type (iv) is the d.c. series motor, in which the input current is also the field current. This means that the back-e.m.f. E in equation (2.1.3) is proportional not only to speed but to *input* current. This in turn means that as the input current and hence the output torque decrease, the speed increases at a greater and greater rate. This can be appreciated by replacing I_f by I in equation (2.1.4) and remembering that T is proportional to I. If the demand on the motor is low enough the motor output speed can thus be excessive and protection is again needed, this time in the form of an overspeed trip.

2.2 General Forms of Demand Characteristic

The demand imposed on a power converter is in the form either of a direct loss (as encountered in overcoming the friction in a pipe line and in a mechanical or electrical resistive device) or of another converter (as in demanding the fluid to drive a turbine, torque to drive a pump or current to drive an electric motor).

Consider first the energy loss incurred in a fluid pipe line. This is determined by the nature of the flow in the pipe line, which usually varies between laminar and turbulent. For the case of laminar flow Newton's law of viscous friction holds in which the shear stress f at the fluid-mechanical interface (the pipe wall) is given by

$$f = \mu dv/dy,$$

where dv/dy is the local velocity gradient across the fluid and μ is the viscosity of the fluid. It may be shown that this leads to the relation

$$Q^n = k_9 P,$$

where n is equal to unity and P is the pressure difference between two specified points along the pipe. The pressure-discharge (P-Q) demand characteristic of the pipe is thus of the form (i) in Fig. 2.2a. As the degree of

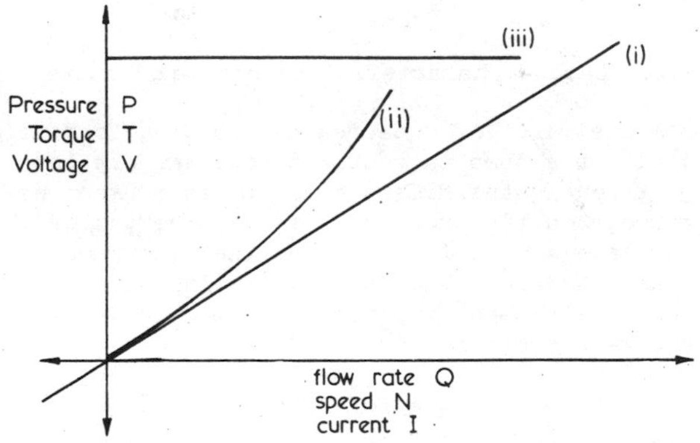

Fig. 2.2a. General forms of demand characteristic

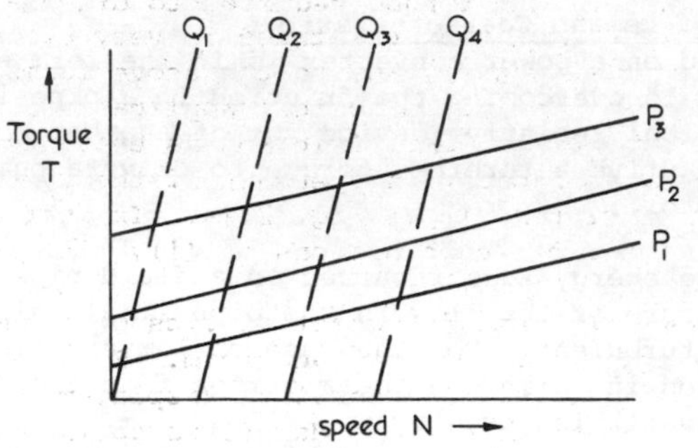

Fig. 2.2b. Demand characteristics of a gear-pump

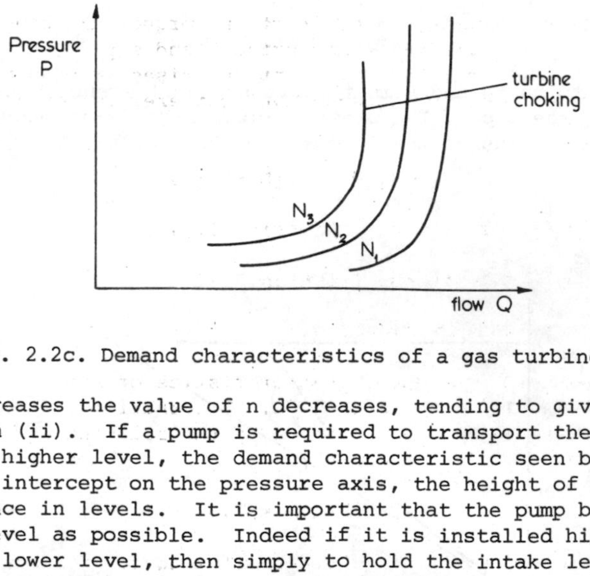

Fig. 2.2c. Demand characteristics of a gas turbine

turbulence increases the value of n decreases, tending to give a characteristic of the form (ii). If a pump is required to transport the fluid from a low level to a higher level, the demand characteristic seen by the pump will incorp.rate an intercept on the pressure axis, the height of which corresponds to the difference in levels. It is important that the pump be placed as near to the lower level as possible. Indeed if it is installed higher than about 10 m above the lower level, then simply to hold the intake leg of fluid stationary will require a suction pressure P_s given by

$$P_s > \rho g \times 10$$

where ρ is the density of the fluid (1000 Kg/m^3 for water) and g is the gravitational constant (9.81 m/s^2). Thus for water we require that

$$P_s > 98.1 \text{ kN/m}^2.$$

Since atmospheric pressure is about this figure then the actual pressure at the pump inlet will be negative. Now negative pressure means that tension will exist in the fluid, causing what is known as separation. Cavities will occur in the fluid and the flow to the pump will be less than expected. If the fluid at the lower level is under a pressure greater than atmospheric then the height of the suction leg can be correspondingly increased without separation occurring.

Consider now the case of the torque required to drive a mechanical resistive device. The origin of the resistance is very often fluid (as in air resistance and bearing resistance), and so the above forms are to be expected and indeed are met a great deal in practical applications. A further form of resistance is due to dry (or coulomb) friction as encountered between dry rubbing surfaces. This is broadly independent of the rubbing velocity and takes the form shown as case (iii) of Fig. 2.2a. In most cases of fluid and mechanical resistance however, combinations of (i), (ii) and (iii) exist.

For the same reasons the demand characteristics of converters (which present a resistance to the supply) are of the same form. Figs. 2.2b and 2.2c show typical demand characteristics for a positive-displacement pump and a gas-turbine respectively. Theoretical considerations applied to the positive-displacement pump indicate that the torque T required to pressurise the fluid to a pressure P can be represented by

$$T = k_{10}P.$$

However, in practice speed-dependent friction torques are caused by viscous drag originating at relatively-moving surfaces and separated by a thin layer of viscous liquid. In addition, dry friction arises at unlubricated surfaces having close fits. The following equation is therefore found to be much more realistic,

$$T = k_{10}P + k_{11}N + k_{12},$$

in which k_{10} accounts for liquid compressibility,

k_{11} accounts for viscous friction

and k_{12} accounts for dry friction.

The constant pressure torque-speed characteristics of Fig. 2.2b are seen to combine curves (i) and (iii) of Fig. 2.2a. Alternatively constant-discharge torque-speed characteristics can be drawn if the relevant equations are known. Here we may use equation (1.3.4) namely,

$$Q = k_5N - k_4P,$$

where k_5 and k_4 are constants, the latter allowing for the small leakage loss resulting from the pressure generated. Thus, for a given value of P, that is, anywhere on a constant-pressure torque-speed characteristic, Q can be found for different values of N. Choosing different values of constant P, a succession of different values of Q can be found, from which constant Q torque-speed characteristics may be plotted (Fig. 2.2b).

Turning now to the turbine demand characteristics of Fig. 2.2c, it will be observed that the constant speed curves run into straight vertical lines at certain values of flow Q. This is caused by flow restriction or choking in the nozzle throats, and also occurs in compressors. The general shape of the curves can be seen to conform with curve (ii) of Fig. 2.2a.

For electrical resistance in the form of transmission loss Ohm's law $V = IR$ holds. Hence the demand characteristic is again of the form (i) shown in Fig. 2.2a. An important feature here is that, for the electrical grid system, the demand may in many cases be assumed to have little effect on the voltage, since the current taken is often relatively small. Hence the demand characteristic is instrumental in determining the current I_1 only (Fig. 2.2d). With a battery supply the situation is totally different and the resulting voltage and current would be given by V_2 and I_2 respectively.

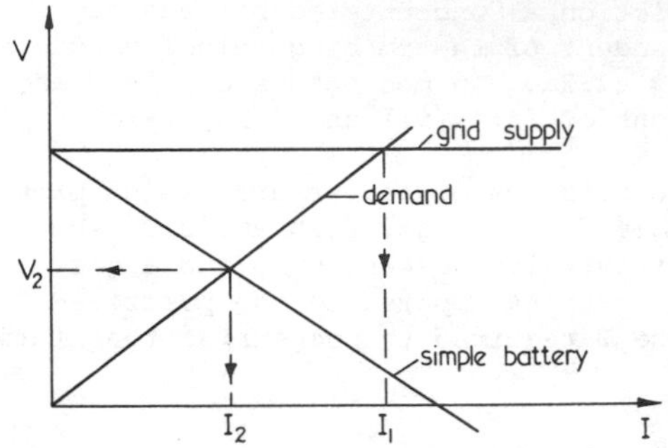

Fig. 2.2d

Just as an electric current encounters resistance so too does heat flow, to varying degrees depending upon the material through which the heat is passing. This is evidenced by the fact that the temperature drops in the direction of heat flow, whether it be by conduction, by radiation or by convection. Conduction occurs through solid materials and is described by Fourier's law,

$$q = K_{th} \, A\Psi/t,$$

where A is the area normal to heat flow, t is the thickness of the material in the direction of flow, Ψ is the temperature difference and k_{th} is the coefficient of thermal conductivity, which is a property of the material. The relation between q and Ψ can thus be seen to be linear. This is not so for radiation and convection. For the former the so-called Stefan-Boltzmann law holds, that is

$$q = k_R \, (\Psi_1{}^4 - \Psi_2{}^4),$$

where Ψ_1 and Ψ_2 are the absolute temperatures of the emitting and receiving surfaces respectively. For convection no generally applicable law holds and each situation must be assessed on its merits, usually experimentally. In the vast majority of cases a combination of conduction, radiation and convection takes place and an overall heat-transfer coefficient is usually defined.

General characteristics of the d.c. machine The basic equations governing the output and demand characteristics of d.c. motors and generators, assuming no losses are

$$V - E \pm I_aR = 0 \qquad (2.2.1)$$
$$E = k_7 \, I_f \, N \text{ volt} \qquad (2.2.2)$$
$$T = k_7 \, I_f \, I_a \text{ Nm.} \qquad (2.2.3)$$

In equation (2.2.1) the positive sign holds for the generator and the negative sign for the motor and

V = input- or output- voltage (motor or generator respectively)

E = back- or generated- e.m.f. (motor or generator)

T = output- or input- torque (motor or generator)

N = output- or input- speed (motor or generator), rad/sec.

I_a = armature current
I_f = field current
R = circuit resistance including armature and series field (if present)
k_7 = machine constant

It is required to set up the torque-speed equation for constant values of voltage and the voltage-current equation for constant values of speed for both the separately excited and the series d.c. machine, when R is 0.25 ohm and 0.55 ohm respectively.

From equations (2.2.1), (2.2.2) and (2.2.3), we may write the torque-speed equation

$$V - k_7 I_f N \pm (T/k_7 I_f)R = 0,$$

and the voltage-current equation

$$V - k_7 I_f N \pm I_a R = 0.$$

Consider first the separately excited machine in which

$$I_a = I, I_f = \text{a constant}$$

and let $k_7 I_f$ = 1.34 volt.s (= 1.34 Nm/A) say, which is fairly typical of a medium-sized machine.

Thus $$V - 1.34 N \pm 0.186 T = 0 \qquad (2.2.4)$$

and $$V - 1.34 N \pm 0.25 I = 0, \qquad (2.2.5)$$

in which the positive and negative signs hold for the generator and motor respectively. Equation (2.2.4) gives the torque-speed characteristics for constant values of voltage V while equation (2.2.5) gives the voltage - current characteristics for constant values of speed N.

Now consider series excitation, in which

$$I_a = I_f = I$$

Thus $$V - k_7 IN \pm 0.55 (T/k_7 I) = 0 \qquad (2.2.6)$$

and $$V - k_7 IN \pm 0.55 I = 0 \qquad (2.2.7)$$

Also from equation (2.2.3)

$$I = (T/k_7)^{\frac{1}{2}} \qquad (2.2.8)$$

A typical value of k_7 in these equations is 0.05 volt.s/A and substitution of this value into equations (2.2.6) and (2.2.7) gives,

$$V + T^{\frac{1}{2}} (\pm 2.45 - 0.223 N) = 0 \qquad (2.2.9)$$

and $$V - 0.05 NI \pm 0.55 I = 0, \qquad (2.2.10)$$

in which the positive and negative signs again hold for the generator and motor respectively. The former is the torque-speed equation and may be plotted for constant values of voltage V; the latter is the voltage - current equation and may be similarly plotted for constant values of speed N. Whether or not such characteristics are realistic depends upon the nature of the drive or of the load. Equations (2.2.4), (2.2.5), (2.2.9) and (2.2.10) are strictly only relevant to an ideal electrical machine in which the magnetic circuit remains unsaturated, that is the field strength continues to increase with increase in field current and with speed. In practice there is a limit to the value the field strength can have and this results in a droop appearing on the voltage-current output characteristics of both the separately-excited and the series-excited generator. In the latter this droop can be sufficient to allow the generator to operate satisfactorily when supplying a load consisting of a low resistance. This feature is often used when series-excited traction motors are run as generators during braking, the braking energy being dissipated in a bank of resistors.

Before leaving the subject of d.c. generators, mention should be made of the shunt-excited generator, in which the field is connected in parallel with the armature circuit. Although this is a self-excited machine it does not suffer from the same disadvantages as does the series generator. This is because the shunt generator can be made to acquire an output voltage *before* connection to the load, that is when the load current is zero, whereas the series generator must be connected to its load before any current will flow in its field windings. When the armature of a shunt generator is rotated there is usually sufficient residual magnetism in the shunt field to enable the generator to create an e.m.f. which in turn drives a current through the shunt winding, thereby creating a greater magnetic field and hence greater e.m.f. and so on. When the electrical load is connected the shunt generator then performs like a separately-excited generator over its usual working range. The droop in the voltage-current output characteristics is however rather more pronounced since a drop in the output voltage results in a fall in the value of field current with a consequent further effect on the output voltage. In fact the stage can be reached when reduction in the output voltage below a certain value can produce a decrease in output current as well as in output voltage and the V-I output characteristic can bend back towards the origin, intersecting the current axis at some positive value.

From equations (2.2.4) and (2.2.6) it may be seen that for a given load torque the speed of both the separately-excited and the series-excited d.c. motor may be increased by reducing the field strength represented by k_7I_f. Physically this may be understood by remembering that a reduction in field strength reduces the back- e.m.f. (equation 2.2.2), allowing an increase in armature current (equation 2.2.1). This causes an increase in torque (equation 2.2.3) which accelerates the armature until the back- e.m.f. has risen sufficiently to reduce the current and torque to a steady value (equations 2.2.1 and 2.2.3). But the motor is now running at a higher speed. Hence weakening the field increases the speed and vice versa. In a shunt motor the field may be weakened by connecting a variable resistance in series with the field winding. This resistance reduces the field current thereby reducing the field strength. In a series motor the field is weakened by connecting a variable resistance in parallel with the field, thus tapping off some of the current without affecting that passed through the armature. Fig. 2.2e shows diagrammatically a series motor fitted with a brake. A voltmeter V is fitted as shown and an ammeter A_1 is placed in one supply line with a variable resistance R_1 in the other. The motor is started by increasing the supply voltage slowly from zero to its full value by adjustment of resistor R_1. The brake is then adjusted and readings taken of speed N, brake torque T, terminal voltage V and supply current I. The test is repeated with the parallel resistance (the diverter) in the circuit and set so that the diverter current is about one third of the total current. Fig. 2.2e shows plots of the torque-speed output characteristic and illustrates the roughly hyperbolic form expected from equation (2.2.9). It can also be seen that the use of the diverter results in a higher speed for a given torque, due to weakening of the field.

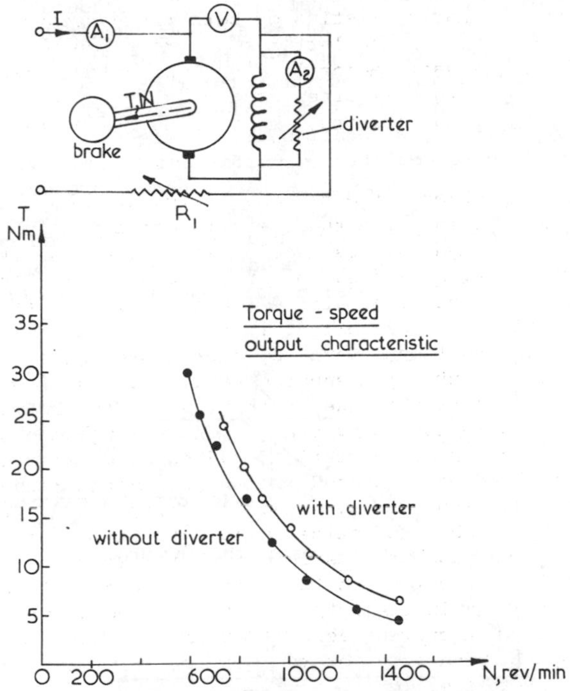

Fig. 2.2e. Use of a diverter

Another important point to be noted from the demand characteristics of d.c. motors as defined by equations (2.2.5) and (2.2.7) is that, at start up or zero speed when the motor is generating no back- e.m.f., the demand current is very large owing to the small value of armature resistance. This is the reason why the series resistance R_1 (Fig. 2.2e) is inserted in the armature circuit. It is gradually cut out as the armature speed and back- e.m.f. increase.

With energy-dissipating loads a limit to the range of performance is often prescribed by the rate of heating and hence by the temperature rise in the load. An obvious case in point is electrical circuitry where the product of demand voltage and demand current, that is the power dissipated, can be high enough to cause damage to the circuit. Another example is the rotodynamic pump operating with its outlet valve closed. If left overlong the heat generated by the impeller would be sufficient to cause damage by overheating. Such a limit is easily appreciated by superimposing upon the family of demand characteristics a rectangular hyperbola whose equation is (in the electrical case)

$$VI = D,$$

where D is the permissible rate of power dissipation. With a d.c. electrical machine the severest limitation is usually in current and we have already seen that, at start-up and at low speed, a resistance is required in series with the armature to limit the starting current.

Fig. 2.2f illustrates this requirement in graphical form using equation (2.2.5) with the negative sign. At a low speed, increasing the series resistance represented by 0.25 in equation (2.2.5) produces a set of matching points such as A, B, C, the last of which is less than the prescribed limiting current, and is therefore safe. When the speed has risen sufficiently the series resistance can be removed and the motor runs safely at point F, say. Now consider the output characteristic of the separately excited d.c. motor (Fig. 2.2g). Owing to its high negative slope care must be taken to ensure that the load demand characteristic is sufficiently low to give a running condition defined by point G, this time less than a prescribed limiting torque, rather than by point H. As mentioned at the end of section 2.1, in case a load having a high-slope characteristic is inadvertently placed on the motor some form of protection is required. This usually takes the form of a fuse or an overload release, both current-sensing devices in the supply to the motor.

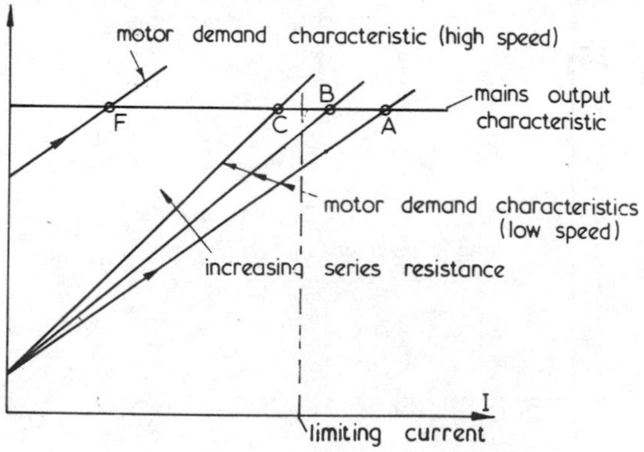

Fig. 2.2f. Demand characteristics of a d.c. motor

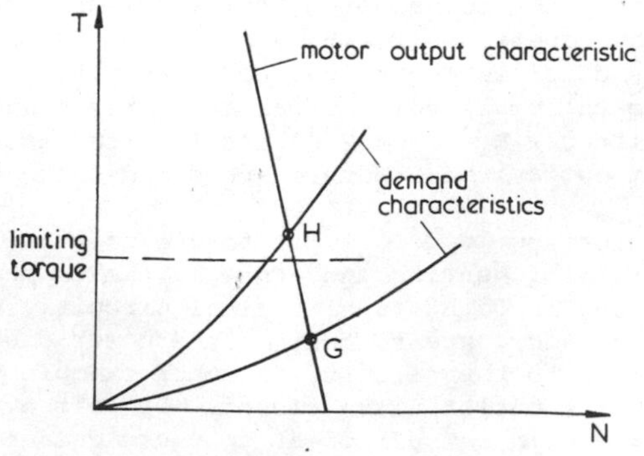

Fig. 2.2g. Working with reference to a limiting torque

Worked example. A separately-excited motor drives a load given by

$$T = AN.$$

Using equations (2.2.4) and (2.2.5) show that, if the load torque is increased at a given speed (i.e. A is increased), then the change in load seen by a constant voltage supply to the motor results in a larger current being supplied to the motor.

Solution. The motor torque-speed output characteristic is given by equation (2.2.4) using the negative sign. Thus

$$0.186\ T + 1.34\ N = V,$$

where
$$T = AN.$$

Hence
$$N = V/(0.186\ A + 1.34).$$

The demand seen by the electrical supply to the motor is given by equation (2.2.5) again using the negative sign. Thus

$$V - 0.25\ I = 1.34\ N = 1.34V/(0.186\ A + 1.34)$$

or
$$V = 0.25\ I\ (1 + 1.34/0.186\ A).$$

It may be observed from this equation that as A is increased a larger current I will be supplied to the motor from a constant voltage supply. In case this current becomes excessive some form of protection will be needed.

This mathematical procedure of obtaining the load characteristic 'seen' by a converter through a second converter has its graphical counterpart in the procedure used in the introductory example in section 1.6.

Worked example. A 250 V d.c. separately excited haulage motor has an armature resistance of 0.05 ohm and an e.m.f. of 245 V at a speed of 1200 rev/min. It is coupled to an overhauling load which imposes a torque of 200 Nm.

Determine the lowest speed at which the motor can hold the load by regenerative braking (that is by the motor operating as a generator).

Solution. Here the load is the supplier of power and the motor constitutes the demand. This is basically a generator problem rather than a motor problem and as such we must use equation (2.2.1) with the positive sign, together with equations (2.2.2) and (2.2.3). Equation (2.2.1) thus becomes,

$$250 - E + 0.05\ I = 0,$$

since
$$I_a = I.$$

Also, equation (2.2.2) applies in which E = 245 V when N = 1200 rev/min.

Thus $k_7 I_f = 245/1200 = 0.205$ volt min/rev

$$= 0.205 \times 60/2\pi = 1.95 \text{ volt.s}$$

$$= 1.95 \text{ Nm/A}.$$

Finally equation (2.2.3) may be applied in which the torque T to be held is 200 Nm. Hence the current generated

$$I = 200/1.95 = 102.5 \text{ A}.$$

Hence $$E = 250 + 0.05\, I = 250 + (0.05 \times 102.5) = 255.125 \text{ V},$$

giving $$N = E/k_7 I_f = 131 \text{ rad/s} = 1250 \text{ rev/min}.$$

This is the lowest speed at which the load can be held.

2.3 General Forms of Efficiency Characteristic

All energy converters suffer from losses of one sort or another. We have already seen in section 1.2 that the mechanical losses occurring in an automotive engine arise from deficiencies in timing and in filling of the cylinders with fuel-air mixture. To these losses can be added those due to friction in the bearings and in any accessories such as oil and water pumps, the fan, the camshaft and the dynamo drive. One of the aims of a converter designer is to ensure that such losses are adequately small at the usual operating point on the converter output characteristic.

Consider the example of the positive-displacement pump already discussed in section 2.2 in which demand torque

$$T = k_{10}P + k_{11}N + k_{12},$$

while from equation (1.3.4) the output flow,

$$Q = k_5 N - k_4 P.$$

Conversion efficiency, η is given by,

$$\eta = PQ/TN$$

and, given the pressure-flow and torque-speed characteristics, it is possible to plot η versus any one of three of the variables depicted on the right-hand side of this equation for a constant value of the fourth variable. The most usual plots are of η versus the output rate parameter (in this case flow Q) or of η versus output power (PQ) for a constant value of one of the input parameters (in this case speed N, say). Assuming k_{12} is negligible, which is a reasonable approximation, it may be deduced that

$$\eta = \frac{Q/N}{k_{10} + k_4 k_{11} \cdot N/(k_5 N - Q)}$$

Thus, for a given value of N the relation between η and Q is almost linear for low values of Q. However as Q increases, η droops more and more from the linear value and curves of the form of Fig. 1.6b usually result. For a medium-sized pump typical values of the constants involved in these equations are as follows,

$$k_5 = k_{10} = 6 \times 10^{-6} \text{ m}^3$$

$$k_{11} = 0.03 \text{ Nms}$$

$$k_4 = 5 \times 10^{-12}, \text{ m}^5/\text{Ns}$$

while k_{12} is often small enough to be neglected.

Similar consideration may be given to the case of the d.c. electrical machine discussed in sections 2.1 and 2.2. For the separately excited motor we may use equation (2.2.3) which, with a value of $k_7 I_f$ of 1.34 gives

$$\eta = TN/V\,I = 1.34\ N/V.$$

For the series motor equations (2.2.8) and (2.2.10) may be used which give

$$\eta = k_7\ N/(0.05N + 0.55).$$

It must be remembered that these equations assume an unsaturated field and thus give optimistic values of efficiency. In electrical machines electrical and magnetic losses occur in addition to mechanical losses. They all appear as heat and raise the temperature of the machine, and indeed the output is limited in most cases by temperature. Electrical machine losses can be divided roughly into two classes, those independent of load such as friction, windage and iron (hysteresis and eddy-current) losses, and those dependent upon load, consisting of winding-resistance losses and brush-contact loss. Fig. 2.3a shows the way in which these losses depend on speed and armature current respectively for a nominal 1 HP (0.746 kW) motor.

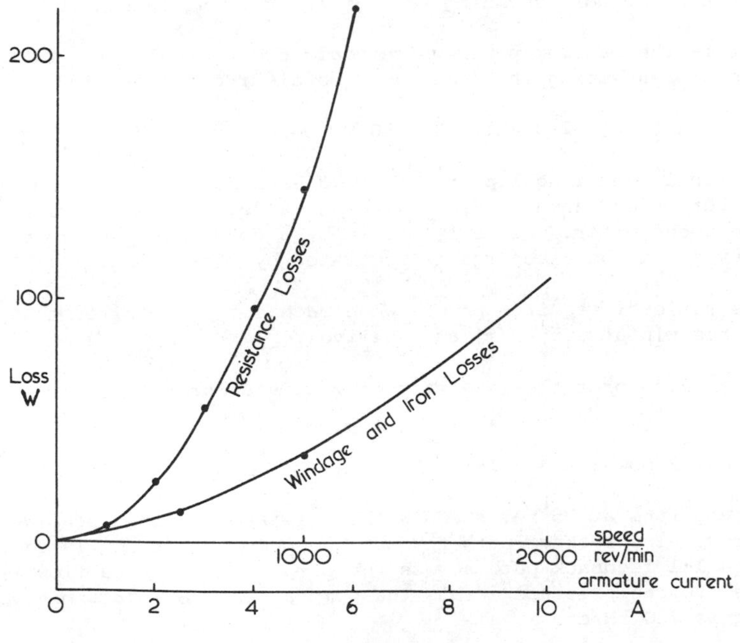

Fig. 2.3a. Electric motor losses

When we are considering a converter producing mechanical power from thermal power, then it is important to remember that there is an upper limit even to its theoretical efficiency. This is dictated by the second law of thermodynamics, and the limiting efficiency η_L is given by

$$\eta_L = (\psi_1 - \psi_2)/\psi_1,$$

where ψ_1 is the absolute source temperature and ψ_2 the absolute sink temperature. No matter how careful we are in the design of an engine its efficiency can never be improved beyond the above limit and so efforts are made to either increase ψ_1 or decrease ψ_2. The former is usually limited by the suitability and strength of materials at high temperatures and the sink temperature is often determined by the particular reservoir used, which is usually a natural one such as the atmosphere or the sea.

2.4 Examples

1. The terminal voltage V and the load current I of a d.c. generator are related by the equation

$$V = 250 - 0.45\ I$$

The generator delivers power to a load resistor of 5 ohms in parallel with a battery on charge, having an internal resistance of 0.33 ohms and an open circuit e.m.f. of 225 volt.

(a) Calculate this power and the current delivered.

(b) If the load is altered such that the load power is to be 14 kW what are the corresponding voltage and current values?

(c) What is the maximum power which could ever be delivered by the generator assuming the load could be altered to take it?

(Ans. (a) 11.6 kW, 51.3 Amp. (b) 220 V, 64 A. (c) 34.75 kW)

2. Two identical centrifugal pump sets A, B each have the P-Q and efficiency-Q characteristics shown in Figs. 2.4a and 2.4b. Set A is driven by a variable-speed motor. Set B is driven by a constant-speed motor running at 15 rev/s and the discharge is regulated by throttling.

Find the ratio of required powers when each set is required to discharge 5000 litres/min at 1.5 kN/m^2 at a delivery point.

Using Fig. 2.4b plot the iso-efficiency curves for $\eta = 0.4$ and 0.6 on Fig. 2.4a.

(Ans. power B/power A = 2.15)

3. A gear-pump runs at 25 rev/s and without leakage would discharge 0.0328 litres/rev. The leakage path through the pump clearances is equivalent to a parallel feedback path across the pump. The leakage flow is proportional to the pressure across the pump and is 28 litres/min when the pressure is 350 kN/m^2.

(a) Draw the pressure-flow characteristic for the pump and find the maximum hydraulic power which can be obtained from the pump.

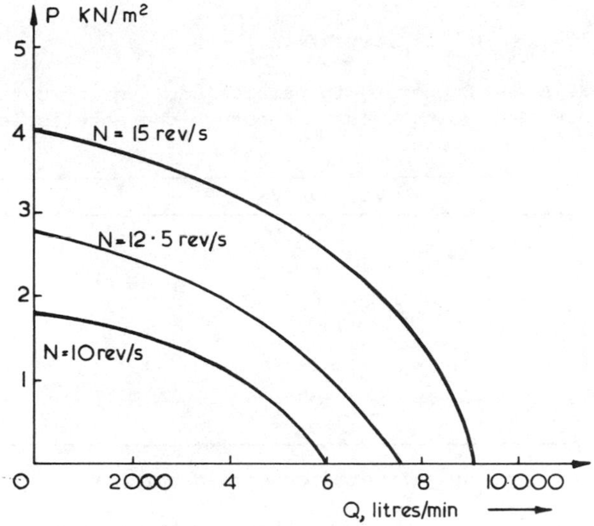

Fig. 2.4a. Pump output characteristics

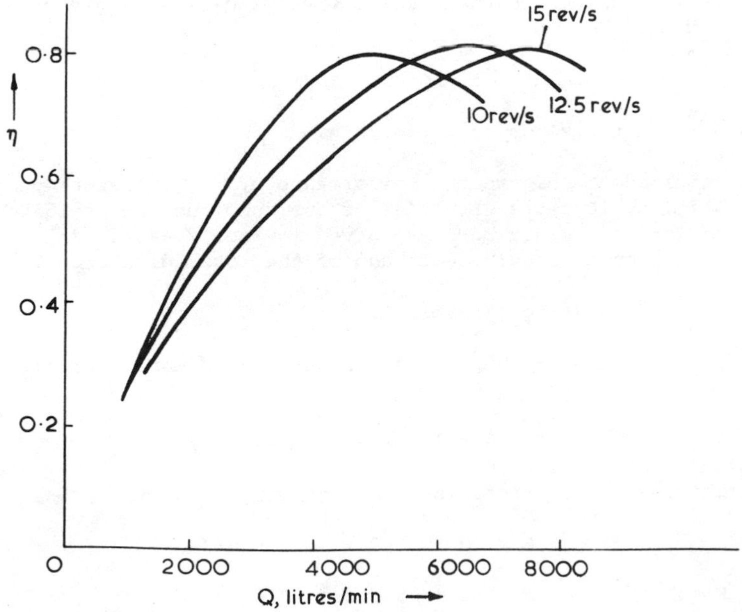

Fig. 2.4b. Pump efficiency characteristics

(b) The pump supplies a system having a resistance proportional to flow and such that 330 kN/m^2 is required to give a flow rate of 45.5 litres/min. Find the flow in the system.

(Ans. (a) 0.126 kW, (b) 31 litres/min.)

4. Two generators A and B operate in parallel delivering current to a 3 ohm resistor. They have voltage-current output characteristics which pass through the following points:

Current A	Voltage V_A	Voltage V_B
0	275	255
12	270	253
25	260	247
33	250	242
42	240	235
48	230	227
55	220	220
61	210	213

(a) Calculate the power delivered by each generator.

(b) If it is required to share the load equally between the generators what must be the load resistance and the output power of each generator?

(Ans. (a) 11 kW, 8.5 kW. (b) 2 ohms, 12.1 kW)

5. The loss (or demand) characteristic of a pipe-line is given in the following equation

$$P = Q^2/16 \quad kN/m^2,$$

where Q is in thousands of litres/min.

It is required to compare the performance of two identical pumps running in parallel at 10 rev/s with that of one pump running at 15 rev/s. The characteristics of each pump are given in Fig. 2.4a. What is the ratio of the total power requirements and of the total discharges?

(Ans. 1.55, 1.23 (single/parallel))

6. A motor drives a gear-pump whose torque-speed demand characteristic is given by

$$T = 6 \times 10^{-6}P + 0.03 \text{ Nm}.$$

The pressure-flow characteristic of the pump is given by

$$Q = 6 \times 10^{-6}N - 5 \times 10^{-12}P \text{ m}^3/\text{s}.$$

where P is in N/m^2 and N is in rad/s.
The pump supplies a piping circuit whose demand on the pump is given by

$$P = 0.6 \times 10^{14} \; Q \quad N/m^2$$

where Q is in m^3/s.

Deduce the equation of the torque-speed demand characteristic seen by the motor.

CHAPTER 3

POWER TRANSMISSION

3.1 The Ratio Element

Consider again the example of the automobile travelling along a road. Let us now find the relations between engine torque T_E and tractive effort E, and between engine speed N_E and forward speed s. These may be obtained with reference to Fig. 3.1a in which the engine supplies the output power $T_E N_E$ needed to overcome the load resistance imposed by the resistive force R on the

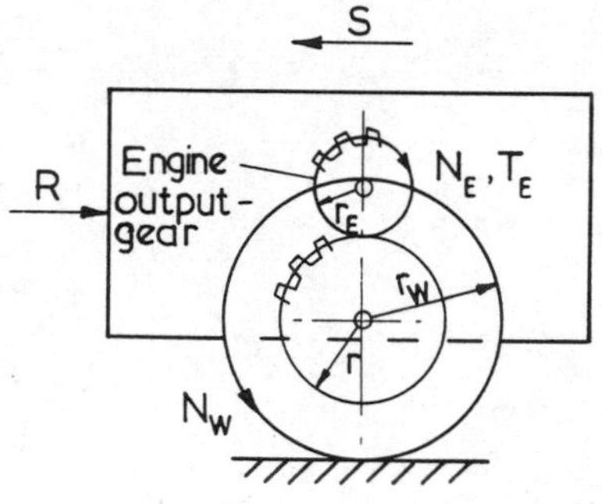

Fig. 3.1a. Gearing in an automobile

vehicle. Let us assume that, in just overcoming R, the engine supplies a forward tractive effort E and drives the automobile forward at constant speed s.

Thus assuming no losses in the transmission the power output of the engine is equal to the power dissipated in overcoming the road resistance R, that is

$$T_E . N_E = R . s$$

However, for a road-wheel diameter $2r_W$, the road-wheel speed N_W is given by

$$N_W = s/r_W.$$

Also the engine shaft is rotating at N_E such that

$$N_E/N_W = r/r_E = G$$

and hence $$N_E = s/Gr_W,$$

where G is termed the speed-ratio or gear-ratio between road-wheel and engine. Hence we may finally write

$$T_E s/r_w G = R.s = E.s.$$

or

$$E = T_E/Gr_w.$$

Thus the output characteristic of the automobile may be represented in axes of tractive effort E and forward speed s as well as in axes of engine torque T_E and engine speed N_E, the simple conversions, assuming no losses being

$$E = T_E/Gr_w$$

and

$$s = N_E.Gr_w \qquad (3.1.1)$$

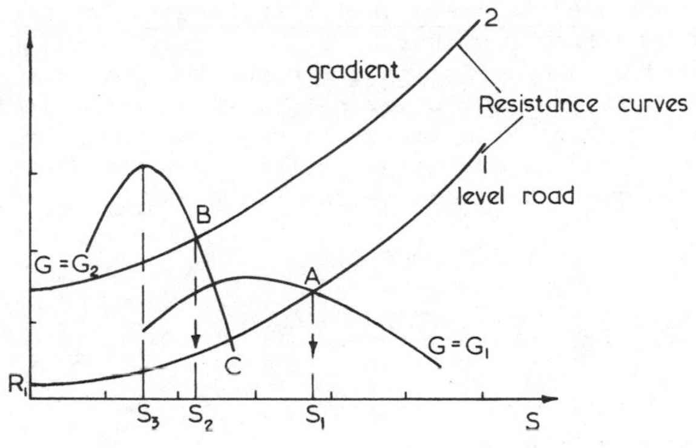

Fig. 3.1b. Matching automobile and road characteristics

The resistance characteristics will be assumed to be of the form shown in Fig. 3.1b, in which curve 2 intersects the vertical axis as a height dependent upon the steepness of the incline up which the vehicle is travelling.

Consider now the torque-speed characteristic of the engine at full-throttle converted to axes of tractive-effort E and forward speed s for a particular gear-box ratio $G = G_1$ (Fig. 3.1b). It is apparent that for this gear ratio the steady forward speed would be given by $s = s_1$, if the car were travelling along a level road. If, however, an incline is encountered, whose resistance characteristic is given by curve 2, then it is clear that the engine can never apply sufficient tractive effort to ascend the incline. The procedure is then to change to a lower gear. Let us assume that the gear ratio G is halved which means that for a given engine torque and speed, the tractive effort is doubled, but the forward speed is halved as can be seen from equations (3.1.1). The new curve is shown at $G = G_2$ (Fig. 3.1b) and the steady forward speed up the incline will be s_2. Hence, by the use of the gear-box (the ratio element), the engine is capable of supplying the required tractive-effort to ascend the incline, albeit at a reduced forward speed. It is quite possible that point B may not correspond to a maximum efficiency point for the engine. To bring B closer to such a point would need a fairly fine adjustment of gear-ratio and it would be simpler to alter the throttle setting so as to operate on a slightly different engine torque-speed curve even though the forward speed might be lower.

When the car is accelerating on a road having no incline (resistance curve 1) and has an instantaneous forward speed s_3 it can be seen that the lower gear G_2 gives a greater excess tractive effort than the upper gear G_1, allowing a greater acceleration up to the point C. This is the reason why a low gear is used for rapid acceleration as in overtaking. To give some idea of practical values, let us consider, say, a car which has a fairly lively performance - a sports coupé weighing about 1 ton laden and powered by an engine developing 100 HP (74.6 kW) at 5000 rev/min. For such a car each interval on the vertical tractive effort axis of Fig. 3.1b would correspond to about 1.3 kN and each interval on the horizontal speed axis to about 25 m.p.h. (40 km/h). G_1 and G_2 would then roughly correspond to top gear and second gear respectively.

Thus we have seen one extremely useful application of the gears embodied in the transmission line from a converter to its load. A slight variant on this application is used when ensuring that a converter operates at its maximum-efficiency point or its maximum-power point. The former is found from the efficiency characteristic. The latter is found by remembering that power is proportional to the product of the potential parameter and the rate parameter, in this case say torque and speed, respectively. Thus for maximum power

$$d(TN)/dN = N.dT/dN + T = 0.$$

Hence

$$dT/T = - dN/N$$

or

$$\log_e T = - \log_e N + \log_e A,$$

where A is a constant of integration.

Hence

$$T = A/N.$$

This is the equation of a rectangular hyperbola. Thus the maximum power point will be the point where any such hyperbola just *touches* the torque-speed characteristic. For curves such as (ii) in Fig. 2.1a an alternative method may be used which relies on the fact that, at the maximum power point

$$dT/dN = - T/N.$$

The line whose slope is -T/N, which just touches the characteristic will thus do so at the maximum power point (Fig. 2.1a). Such a point is best determined by laying a ruler at a tangent to the output characteristic so that AC = CO and OD = DB. Then

$$- dT/dN = AO/OB = CO/OD = T/N.$$

The automotive gearbox discussed above can be contrasted with a marine gearbox in which a small gear or pinion connected to the engine drives a final larger gear connected to the propeller shaft. Since a ship does not have to climb hills like a car and is rarely called upon to tow another vessel (when a high torque at low speed might be necessary), the gear ratio is not usually made adjustable except perhaps to give reverse and neutral if the prime mover is not reversible. Instead it is simply designed so that, under normal running conditions, the engine is operating at its maximum power point. However, when a ship is towing and its forward speed is reduced the propeller is not at its most efficient, having been designed for operating at the higher forward speed. Rather than employ a variable-speed gearbox to give a lower propeller speed and hence restore its efficiency, such efficiency is often enhanced at the higher propeller speed by varying the blade-pitch. Indeed extreme variation of the blade-pitch is sometimes the means by which a ship can be made to run astern.)

Worked Example. A water-wheel has a torque-speed output characteristic which passes through the following points:

Speed, N_w rev/s	0	1	2	3	4	5
Torque, T_w Nm	140	140	135	115	70	0

(a) It is desired to utilise the maximum power output. At what speed should it be run?

(b) The water-wheel drives a mill whose torque-speed demand characteristic is given by

$$T_m = 70\ N_m^2 \text{ Newton-metres}$$

where N_m is the mill speed in rev/s. Calculate the gear-ratio, $G(=N_m/N_w)$ required to obtain maximum power at the mill, assuming no transmission losses.

(c) Plot the torque-speed demand characteristic 'seen' by the water-wheel for the gear-ratio determined above.

Solution (a) Fig. 3.1c shows the torque-speed output characteristic of the water-wheel, from which the maximum power is obtained by the procedure outlined previously. This occurs at

$$N_w = 3 \text{ rev/s}$$

and

$$T_w = 115 \text{ Nm.}$$

Hence maximum power = 115 x 3 x 2π = 2168 W.

(b) Power extracted by the mill = $T_m N_m = 70\ N_m^2 \times 2\pi\ N_m$ and this is to equal the maximum power delivered by the water-wheel. Thus

$$70 \times 2\pi \times N_m^3 = 2168,$$

whence

$$N_m = 1.7 \text{ rev/s.}$$

But

$$N_w = 3 \quad \text{rev/s}$$

and hence

$$G = N_m/N_w = 0.567.$$

(c) The equation of torque-speed demand characteristic seen by the water-wheel is

$$T_{mw} = 0.567\ T_m = 0.567 \times 70\ (0.567\ N_w)^2$$

or

$$T_{mw} = 12.76\ N_w^{\ 2}.$$

This is plotted on Fig. 3.1c and is observed to intersect the output characteristic of the water-wheel at it maximum power point.

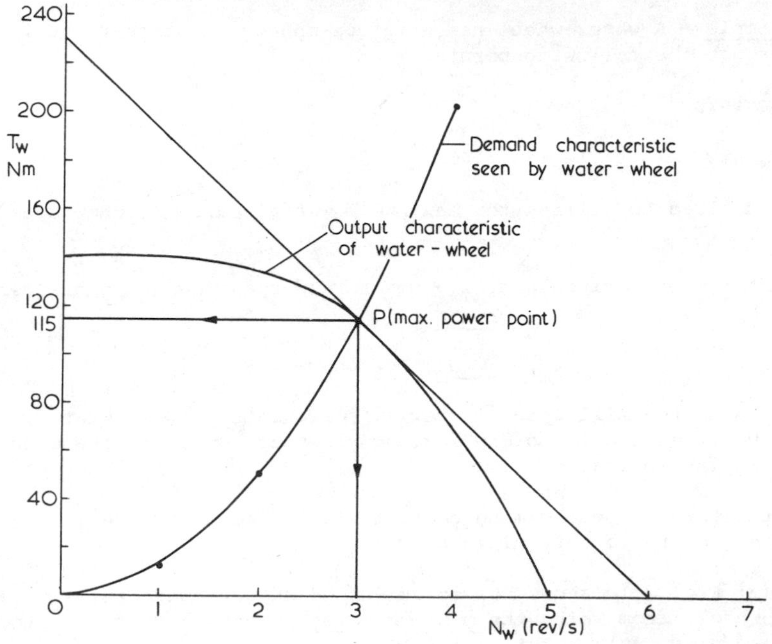

Fig. 3.1c. Matching a water-wheel to its load

The question now remains as to how we design a gear-train or gear-box to provide a certain prescribed gear-ratio G between output and input speeds. Such gear-trains may be sub-divided into two types - simple and compound, and Fig. 3.1d exemplifies these. In the upper diagram there is only one gear on each shaft and in the lower a pair of integral gears is used on one or more shafts. For the former,

$$G = \frac{\text{No. of teeth on input gearwheel}}{\text{No. of teeth on output gearwheel}},$$

and the intermediate gears are known as idlers, being used to bridge the gap between the input and output shafts. The automotive gearing of Fig. 3.1d must be so arranged that the camshaft revolves at half the crankshaft speed, and so wheel D must have twice as many teeth as has wheel A. The drive to the dynamo is taken from wheel C and, as this has 45 teeth, the dynamo pulley is driven at two-thirds the crankshaft speed.

For the compound gear-train shown in Fig. 3.1d, gear-ratio

$$G = (M/A).(B/H) = (10/30).(10/40) = 1/12,$$

where the capital letters denote numbers of teeth. The above value of G is the required gear ratio between the speed of the hour hand of the clock and the speed of the minute hand, the former being driven from the latter.

The distance between successive teeth on a gearwheel is known as the circumferential pitch and is the same value for all gear wheels which mesh together. When the radius of the larger gear is made inifinite it is termed

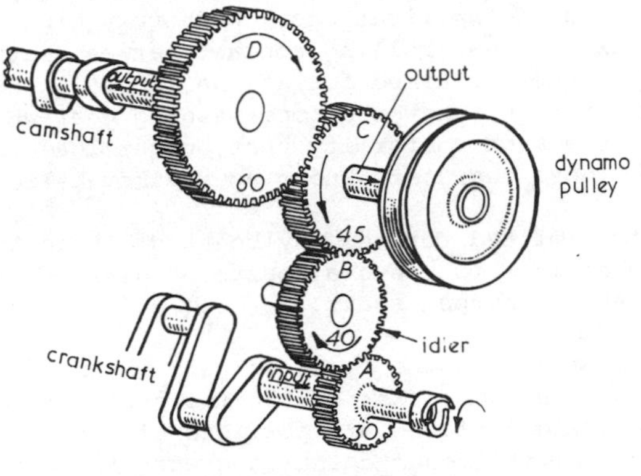

Simple gear train used in automotive accessory drives

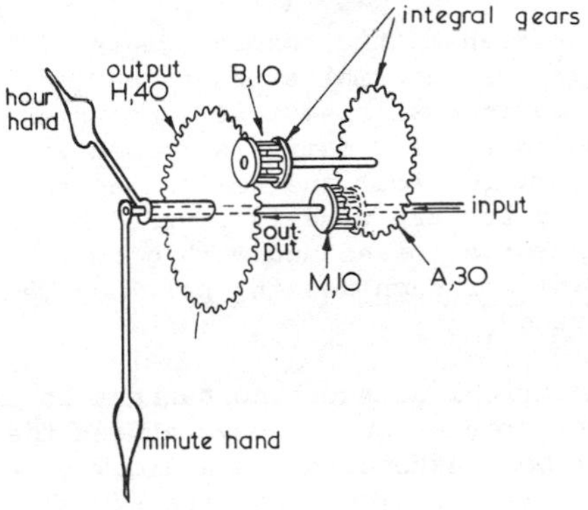

Compound gear train used in clock hand - drive

Fig. 3.1d. Types of gear-train

a rack the smaller gear being called a pinion. The gears depicted in Fig. 3.1d are all externally meshing. However, the same analysis applies if the gears mesh internally, the larger gear of a meshing pair having internal teeth and the smaller, external. An advantage in using such a combination is that a more compact arrangement results. The gearwheels of a meshing pair now rotate in the same directions, whereas with external meshing they rotate in opposite directions.

(The shape of the teeth on a gear-wheel demands consideration. Mating teeth must constitute what is known as a conjugate pair, that is a pair in pure rolling contact. The most common shapes are those of the involute and the cycloid, for which there are standard specifications. When the flank faces of the teeth are parellel to the shaft axes the gears are known as spur gears but these tend to be rather noisy at high speeds and so are usually restricted to low speed open-gear applications. With heavily loaded gears it is necessary to arrange for the tooth loads to be taken up gradually so that, at a given time, several pairs of teeth may be in contact. A way of achieving this is to dispose each gear tooth on a helix around the cylindrical surface of the gear-wheel, which then becomes known as a helical gear. Such gears are used extensively in marine applications. With very high loads, distortion occurs to such an extent that rolling contact between mating teeth cannot be achieved and area contact is aimed for at the outset. The result is the circular arc tooth profile, convex on one meshing gear-wheel and concave on the other, which gives such conformity that, under load, area contact is readily achieved. These gears are known as Novikov-helical gears.)

(In some cases transmitted torques may be so great as to necessitate the use of split-power-trains in which several pairs of gears share the power before it is re-combined at the output shaft.)

(Also worthy of mention are the types of gear which transmit motion between non-parallel shafts. When the axes of the two shafts intersect, bevel gears are used and the driving surfaces are portions of cones instead of cylinders, the imaginary apexes of the cones coinciding with the point of intersection of the shafts. When the shaft axes are skew, that is non-intersecting and non-parallel, skew gearing is used.)

The power that can be transmitted through gears of differing types is limited by considerations of material strength and the method of lubrication. The largest powers and peripheral speeds are encountered in helical gears and can be as much as 37000 kW and 150 metre /s respectively. The largest ratios normally met are about 8:1 for spur and helical gears and about 5:1 for bevel gears. Power losses in gears are usually divided between tooth-friction losses and lubricant churning losses and with good lubrication, losses of efficiency as low as 0.5% for each meshing pair may be expected and indeed are obtained in marine gearing.

The capacity of electrical alternating current to produce mechanical rotation through a converter creates an analogy between the electrical transformer and the mechanical gear box, extending the analogy between the d.c. potentiometer and the mechanical lever. The ideal transformer transmits a.c. power from one part of an electrical system to another, such that both voltage and current are changed, but the power remains constant. Transformers consist of two coils wound on a common magnetic core, the input coil being known as the primary and the output the secondary. If the primary coil p has n times as many turns as the secondary coil s then we may say that the current ratio,

$$I_s/I_p = n$$

and the voltage ratio,

$$V_s/V_p = 1/n,$$

in which n is analogous to the speed ratio of a gear-box.

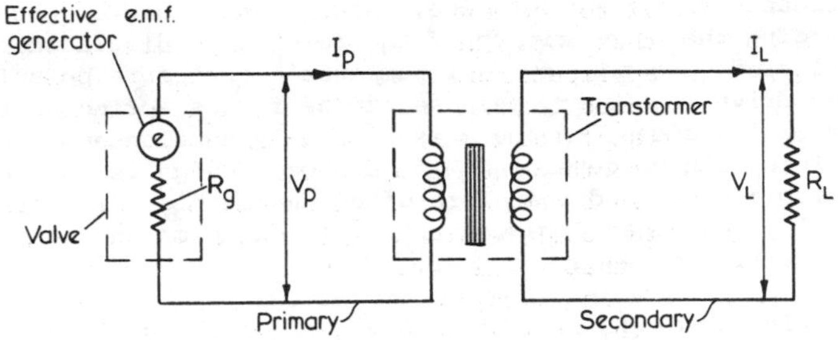

Fig. 3.1e. Equivalent circuit of an electronic valve and its load

Worked example. An a.c. power source consists of an electronic valve which delivers power to a load resistance R_L of 16 ohm through a transformer. The equivalent circuit is shown in Fig. 3.1e in which the internal resistance R_g of the valve is 40,000 ohm. Deduce the turns ratio, primary to secondary, of the transformer so that the power delivered to the load is a maximum.

Solution. For the secondary circuit

$$R_L = V_L/I_L = (V_p/n)/nI_p$$
$$= V_p/n^2I_p,$$

where n is the turns ratio.

Thus the effective load resistance R seen by the generator is given by

$$R = V_p/I_p = R_L n^2.$$

Hence the total effective resistance in the primary circuit is $(R + R_g)$ and the power P supplied to the load is given by

$$P = I_p{}^2 R$$
$$= e^2R/(R + R_g)^2.$$

This is a maximum when dP/dR is zero, that is when $R = R_g$ or when $R_L = R_g/n^2$.

Hence $$n = (R_g/R_L)^{\frac{1}{2}} = (40{,}000/16)^{\frac{1}{2}} = 50.$$

A version of the gearbox in which mechanical power is transmitted with the aid of fluid power is the hydrostatic transmission which consists of a positive-displacement hydraulic pump such as a gear-pump, driven by the converter, feeding oil under pressure to a hydraulic motor, which acts in the reverse way to a positive-displacement pump and drives the load.

Other pumps falling into the same positive-displacement category are vane pumps and piston pumps. In the former the moving member consists of a rotor in which vanes are carried in radial slots. The tips of the vanes are forced against the stationary casing either by springs or by hydraulic pressure. The rotor is offset from the stator centre so that, as the rotor turns, the space between the two increases on the inlet side, sucking fluid into the pump, and decreases on the outlet side, forcing the fluid out of the outlet port. By changing the eccentricity of the rotor, the displacement of the pump can be altered, and this type of pump would therefore be suitable for a variable speed drive. However, because of the forces acting on the rotor at the inlet and outlet sides, the pump is basically unbalanced and this limits the pressure which can be developed. A balanced design is possible, but this is of fixed displacement and therefore often unsuitable. The slip in a vane pump is generally less than in a conventional gear pump and hence volumetric and mechanical efficiencies are higher.

In the positive displacement piston pump, the 'casing' is a cylinder block and the moving members are pistons sliding in the cylinder bores. The pumping action is obtained by reciprocating the pistons relative to their bores and feeding the fluid to and from the cylinders by inlet and discharge valves. There are two basic types of piston pump - radial and axial. In the radial pump the pistons are set radially about the driving shaft and pumping action is obtained either by reciprocating the pistons in a stationary block or by moving the block relative to the pistons. In axial pumps, the reciprocating action is obtained either by setting a swash plate at an angle to the pump axis or by tilting the cylinder block itself. A swash-plate is a plate mounted on the shaft at a set inclination (Fig. 3.1f). During pump-shaft rotation a ring of plungers is sequentially pushed in to dispatch oil into the system and an oil reservoir is usually employed at the pump outlet to smooth

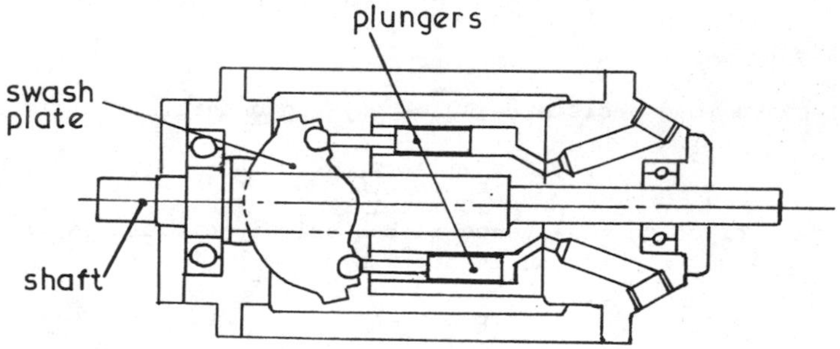

Fig. 3.1f. Swash-plate pump

out any pulsations in the flow. Both radial and axial piston pumps lend themselves very readily to design as variable output units. In the radial pump the output is varied by altering the degree of eccentricity of the cylinder block. With axial pumps, the degree of tilt of the swash plate or the cylinder block determines the output of the pump. Volumetric and mechanical efficiencies of piston pumps are usually better than for vane pumps, because of the good sealing possible between the pistons and their bores and this means that they can generate much higher pressures.

It has already been shown in section 1.3 that the discharge Q from a positive-displacement pump is given by equation (1.3.4). Here let

$$Q = k_5 N_i - k_4 P, \tag{3.1.2}$$

where N_i is the input speed and k_4P is unavoidable leakage within the pump which detracts from the overall pump delivery. In a similar way a leakage flow Q_L occurs in a hydraulic motor such that its output speed N_o is given by

$$N_o = (Q - Q_L) k_{13}.$$

The leakage flow Q_L is dependent upon the supply pressure P and the above equation becomes

$$N_o = (Q - k_{14}P)k_{13}, \tag{3.1.3}$$

in which k_{13} and k_{14} are constants. As in the case of a gear pump, speed-dependent frictional losses exist in a hydraulic motor, which require some of the supply pressure P to overcome them. Thus the motor output torque T_o may to a fair approximation be written,

$$T_o = k_{15} P - k_{16} N_o \tag{3.1.4}$$

Using equations (3.1.2) to (3.1.4) it can be readily shown that the torque-speed output equation of a hydrostatic transmission is given by

$$T_o = \frac{k_5 k_{15} N_i}{k_4 + k_{14}} - N_o \left(\frac{k_{15}}{k_{13} (k_4 + k_{14})} + k_{16} \right) \tag{3.1.5}$$

It is obvious from equation (3.1.5) that for a given pump input speed N_i and a given demand characteristic the motor output speed can be altered by varying the leakage constants k_4 and/or k_{14} or by varying the constants k_5 and/or k_{13}. k_4 and k_{14} can be varied by employing a return bypass from outlet to inlet with an adjustable valve. Fig. 3.1g shows a simple hydrostatic drive using

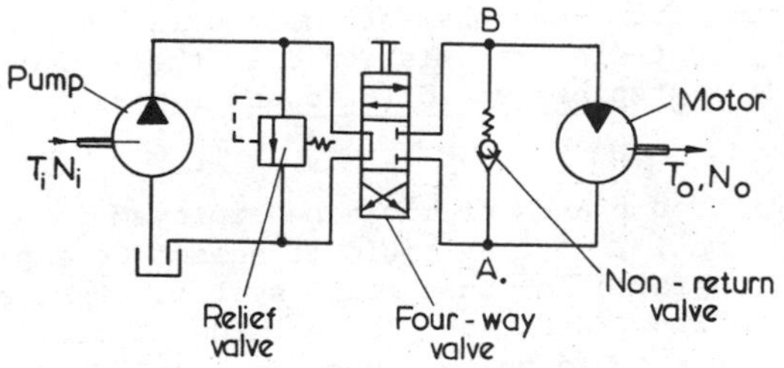

Fig. 3.1g. Simple hydrostatic transmission

the conventional symbols for the components. It consists of a fixed-displacement pump and a fixed-displacement motor. Such a system is used in many hydrostatic drives of limited power of up to about 40 kW, such as fan drives, conveyor drives and auxiliary marine drives. In case the type of hydraulic motor is one which can only operate satisfactorily in one specific direction it is necessary to provide the non-return valve shown in Fig. 3.1g. This ensures that if the flow is in the wrong direction, it will be short-circuited from A back to B. The speed and direction of the motor are normally varied by spilling off the pump flow within the four-way valve shown.

For large drives the above method is very wasteful of power and a common way of varying k_5 and k_{13} is to use a swash plate on both pump and motor. Fig. 3.1h shows a set of typical performance characteristics for a small

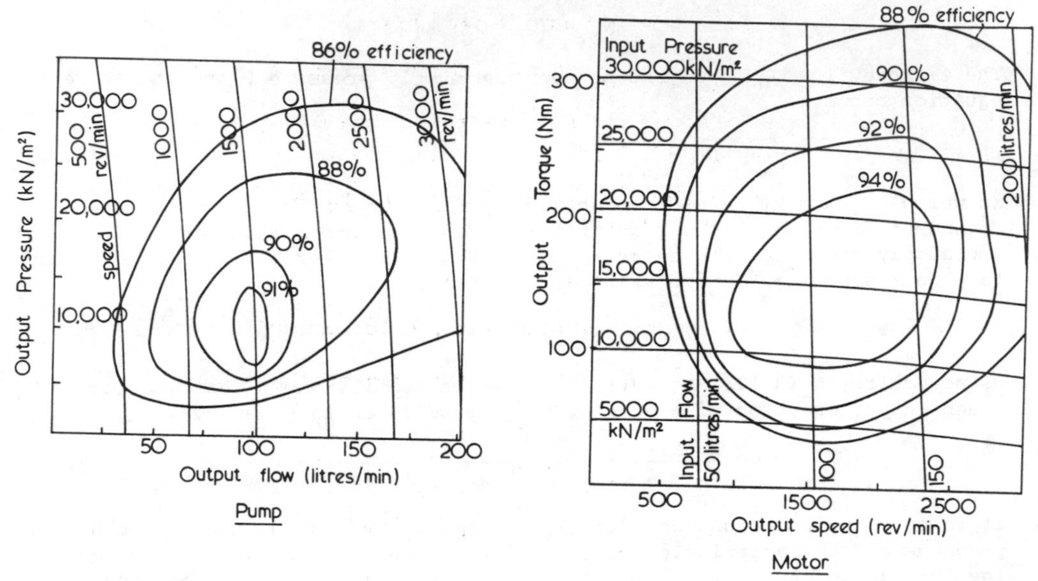

Fig. 3.1h. Hydraulic pump and motor characteristics

hydraulic pump and a small hydraulic motor, both employing swash-plates. The loop form of the constant-efficiency lines should be noted. Such a form is deducible for the pump from the constants quoted in Section 2.3. Two obvious advantages of the hydrostatic transmission over the mechanical gearbox are its abilities to provide a stepless speed ratio and a positive torque at zero load-speed.

Just as a hydraulic pump and motor can be employed in a mechanical system to give a variable gear-ratio, so they could if needed be employed, in the reverse order, to provide transformation in a fluid system. Thus a fluid source of high pressure and low flow rate, say, can be made to drive a hydraulic swash-plate motor. This motor could be mechanically coupled to a hydraulic pump, from which can be extracted fluid power at say low pressure and high flow rate.

Worked example. A transmission system consists of a positive-displacement pump and a positive-displacement motor. The discharge Q from the pump is given by equation (3.1.2),

$$Q = k_5N_i - k_4P.$$

The flow demand Q of the motor is given by equation (3.1.3),

$$Q = N_o/k_{13} + k_{14}P.$$

The motor output torque T_o is given by equation (3.1.4),

$$T_o = k_{15}P - k_{16} N_o.$$

The pump demand torque T_i is given by

$$T_i = k_{10} P + k_{11} N_i.$$

In these equations, N denotes speed and P denotes pressure.

(a) Deduce the forms of the demand torque and of the output torque of the transmission in terms of N_i and N_o.

(b) The transmission is used to couple a converter to its load whose characteristic is given by

$$T = K_6 + K_7 N.$$

Deduce the equation of the load characteristic seen by the converter.

Solution. (a) The output torque, T_o from the transmission is given by

$$T_o = k_{15} P - k_{16} N_o$$

in which $$P = (Q - N_o/k_{13})/k_{14}.$$

But for pump $$Q = k_5 N_i - k_4 P$$

and thus $$Q = k_5 N_i - (k_4/k_{14})(Q - N_o/k_{13}),$$

whence

$$Q = (k_5 N_i + N_o k_4/k_{13} k_{14})/(1 + k_4/k_{14}).$$

This gives $P = (k_5 N_i + N_o k_4/k_{13} k_{14})/(k_{14} + k_4) - N_o/k_{13} k_{14}$,

whence $$T_o = k_{15} \left[\frac{\left(k_5 N_i + \dfrac{k_4}{k_{13} k_{14}} N_o\right)}{(k_{14} + k_4)} - \frac{N_o}{k_{13} k_{14}} \right] - k_{16} N_o.$$

This is of the form

$$T_o = K_1 N_i - K_2 N_o.$$

The demand of the pump is given by

$$T_i = k_{10} P + k_{11} N_i.$$

Substituting again for P gives the form

$$T_i = K_3 N_i - K_4 N_o.$$

(b) Consider the output connection from the transmission to the load. Thus at the equilibrium operating speed

$$K_1 N_i - K_2 N_o = K_6 + K_7 N_o,$$

whence
$$N_o = \frac{K_1 N_i - K_6}{K_2 + K_7}.$$

Substitution then gives

$$T_i = K_3 N_i - K_4(K_1 N_i - K_6)/(K_2 + K_7)$$

or
$$T_i = N_i \left|K_3 - K_4 K_1/(K_2 + K_7)\right| + K_4 K_6/(K_2 + K_7).$$

This illustrates a demand characteristic whose intercept on the torque axis is given by

$$K_4 K_6/(K_2 + K_7)$$

in contrast with K_6 for the load by itself. By adjustment of K_2 and K_4 this intercept may be modified as required and also the demand characteristic may be made to intersect the converter characteristic at the maximum power point.

An electro-mechanical version of the hydrostatic drive is the so-called Ward-Leonard drive which is used in such applications as machine tools, rolling mills and lifting gear. It consists of a d.c. generator supplying current to a d.c. motor, both separately excited, the diagrammatic layout being shown in Fig. 3.1i.

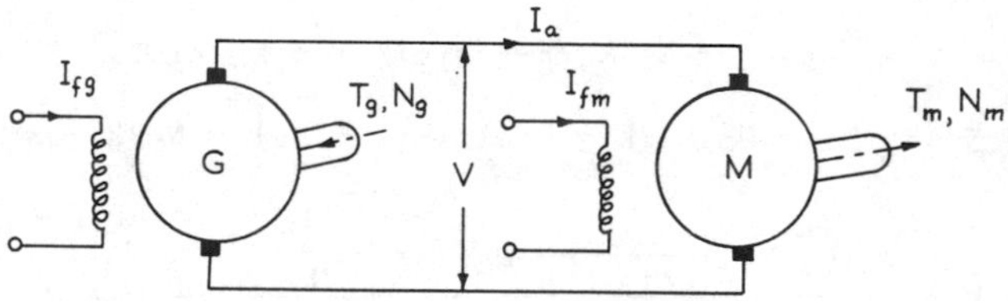

Fig. 3.1i. Ward-Leonard speed control

From section 2.2 it has been shown that, for both motor and generator (equations (2.2.1) to (2.2.3)),

$$V - E \pm I_a R = 0$$

$$E = k_7 I_f N$$

and
$$T = k_7 I_f I_a,$$

from which
$$V - k_7 I_f N \pm I_a R = 0$$

Thus, for the generator G of the Ward-Leonard drive

$$V = k_{7g} I_{fg} N_g - I_a R_g,$$

while for the motor M

$$N_m = E_m/k_{7m}\, I_{fm} = (V - I_a\, R_m)/k_{7m}\, I_{fm}.$$

Hence $$N_m = N_g \quad k_{7g}\, I_{fg}/k_{7m}\, I_{fm} - I_a(R_g + R_m)/k_{7m}\, I_{fm}$$

Thus, for a given generator input speed N_g, N_m may be varied by I_{fg} and/or I_{fm}. Since the motor output torque T_m is given by

$$T_m = k_{7m}\, I_{fm}\, I_a,$$

then the output torque-speed equation for the Ward-Leonard drive is given by

$$N_m = (k_{7g}\, I_{fg}/k_{7m}\, I_{fm})\, N_g - T_m\, (R_g + R_m)/k_{7m}{}^2\, I_{fm}{}^2 \qquad (3.1.6)$$

Equation (3.1.6) may be rewritten in the form

$$T_m = A\, N_g - B\, N_m$$

and as such is directly comparable with equation (3.1.5). Thus at zero motor speed a positive output torque exists, making the system suitable for locomotives and automobiles as an alternative to a gearbox plus clutch or other ratioing plus slipping device. Unfortunately this drive tends to be rather bulky and costly and so has dropped out of favour for automobiles, although in the 1920's so-called petrol-electric commercial vehicles were produced with very good operating characteristics and having the advantage that their 'gear-ratio' was continuously variable.

3.2 The Slipping Element

The main problems associated with the torque-speed characteristics of form (i) (Fig. 2.1a) arise from the fact that at zero speed the output torque is also zero. This means that some form of independent starting is needed, usually by the application of an *external* torque. Also some form of temporary load decoupling is needed to allow an adequate speed and hence torque to be generated by a converter before it is connected to the load which perhaps demands a substantial torque at zero load-speed. Such a load occurs when a vehicle is required to start on a hill, say.

The simplest device for decoupling a converter from its load is the clutch used in many vehicle drives. The eventual connection to the load involves a slipping phase in which the speeds of the load and of the engine are allowed to gradually approach each other.

Another obvious use of the clutch is in changing gear. A gear change can only be carried out when the gearbox is temporarily disconnected from the engine, since the meshing pair forming the path of power flow must be changed. The clutch is the obvious point at which such a disconnection can be effected and so it is invariably mounted between the engine and the gearbox. A typical clutch is shown in Fig. 3.2a and consists of three discs: the flywheel attached to the output shaft of the engine, the clutch-plate and the pressure-plate. After the correct gear has been selected the springs are allowed to push the clutch-plate against the flywheel, and they first rub together during slip then turn together, sending power to the driving-wheels. There is a variety of clutches on the market utilising different physical principles. The magnetic-particle clutch is controlled by an external magnetic field which lines up magnetic particles to form a magnetic connection between input and output

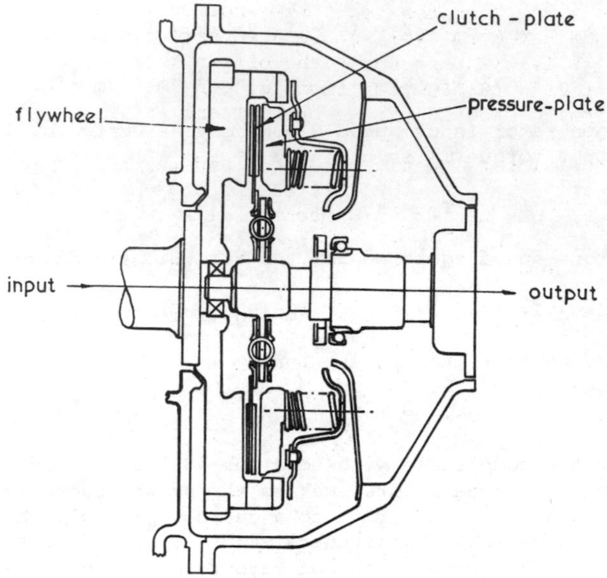

Fig. 3.2a. Automotive clutch

shafts. The liquid-slip clutch has plates whose relative clearances may be adjusted to provide a fluid connection of various strengths. The electro-viscous clutch is controlled by an external field which influences the viscosity of fluid interposed between input and output plates. All may be used as variable speed drives of limited capacity.

A more sophisticated form of the clutch is the fluid coupling which is usually depicted diagrammatically as shown in Fig. 3.2b. The two parts of the

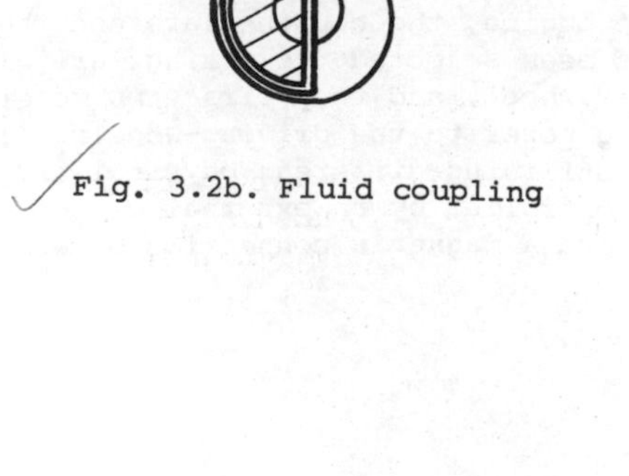

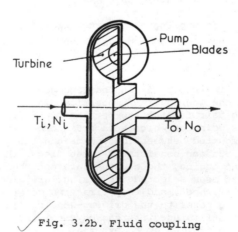

Fig. 3.2b. Fluid coupling

coupling rotate together, the pump portion with the prime mover or other drive converter and the turbine with the load, the whole unit being nearly full of oil. Both pump and turbine units resemble a saucer with a number of flat radial blades. Power is transmitted to the oil as kinetic energy, and this oil is then passed to the turbine for reconversion to mechanical power. As there is no mechanical connection between input and output, there is a certain flexibility which is not present in a gear drive or hydrostatic transmission. Both pump and turbine slip to a certain extent and the heat generated can be conveniently extracted by means of an oil cooler and not left to cause breakdown by overheating as can happen in a mechanical clutch which is caused to slip. The cushion of oil also serves to protect the converter from any shock loads imposed by the demand. Because of slip the turbine runs slower than the pump,

that is $$N_o < N_i.$$

The transmission efficiency η is given by

$$\eta = T_o N_o / T_i N_i,$$

while the relative slip is defined as

$$s = (N_i - N_o)/N_i.$$

These equations thus give

$$\eta = (1 - s)\, T_o/T_i.$$

The same variables apply to this device as apply to the centrifugal pump and turbine, since the latter are embodied in the device. The 'pump' has a torque-speed demand and a pressure-flow output; the 'turbine' has a pressure-flow demand and a torque-speed output and both elements contribute towards the vital 'slip' feature. Non-dimensional groups which govern the performance may be obtained from a knowledge of the important variables, which in this case are

$$T_i,\ D,\ N_i,\ N_o,\ \rho \text{ and } \mu,$$

where D is the coupling diameter and ρ and μ are the fluid density and viscosity, respectively. Applying the rules of dimensional analysis we obtain,

$$T_i/\rho N_i^2 D^5 = f(N_o/N_i,\ \rho N_i D^2/\mu),$$

where $\rho N_i D^2/\mu$ is termed the Reynolds Number, its effect in pumps and couplings being usually very small. Since slip s is a function of speed ratio N_o/N_i, then it may be seen that

$$T_i/\rho N_i^2 D^5 = \Psi\,(s).$$

In fact the particular function is of the form,

$$T_i/\rho N_i^2 D^5 = k_{16}\,\left[1 - (1 - s)^4\right], \qquad (3.2.1)$$

where k_{16} is a constant, dependent on the amount of oil supplied to the unit. Thus for a given slip s, T_i is proportional to N_i^2, and the demand characteristic of the fluid-coupling takes the form shown in Fig. 3.2c. To find steady-state values of T_i and N_i in a given application the relevant curve must be matched to the output torque-speed curve of the appropriate converter, say an automotive engine (i) or an induction motor (ii).

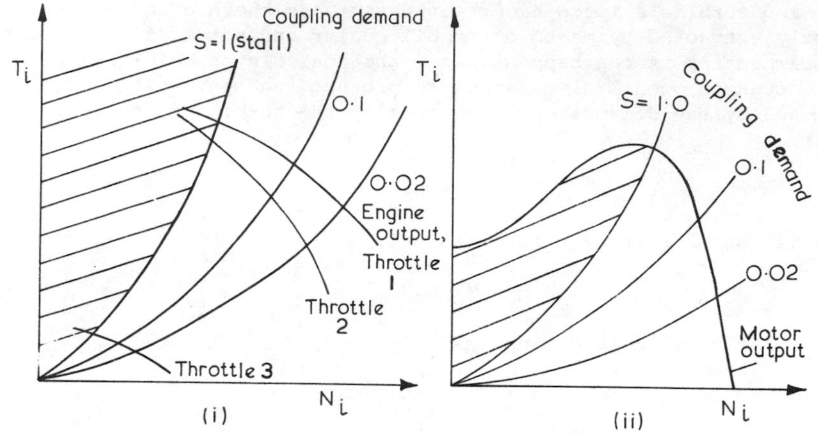

Fig. 3.2c. Matching a fluid coupling to a converter

Consider now the output characteristic of the fluid coupling. Under steady state conditions it may be assumed that $T_o = T_i$. Then using equation (3.2.1),

$$T_o/\rho N_i^2 D^5 = k_{16} \; |1 - (1 - s)^4|.$$

Now

$$1 - s = N_o/N_i,$$

and thus

$$T_o/\rho N_o^2 D^5 = k_{16} \; (N_i/N_o)^2 \; |1 - (N_o/N_i)^4|$$

and

$$T_o = \rho D^5 \, k_{16} \; |N_i^2 - N_o^4/N_i^2|. \qquad (3.2.2)$$

Thus an output torque is generated when the output speed is zero, which is characteristic of a slipping device. The constant k_{16} again depends on the amount of oil supplied to the unit which can often be manually controlled in order to produce at a given load a given output speed. Also a load can be driven whose torque-speed demand characteristic may have an intercept on the torque axis (Fig. 3.2d) even though the drive converter (say an engine) may exhibit no such intercept. As in the case of the clutch this is conditional upon the load being temporarily held until the value of N_i has risen sufficiently. In the case of a car engine the torque intercept is often due to a gradient and the means by which it is held is the handbrake.

The electrical counterpart of the fluid-coupling is the eddy-current coupling which uses an electromagnetic field for the transmission of power rather than a fluid field and ratings ranging from less than 100 kW to 2 MW are typical. The elementary form of a practical eddy-current coupling comprises a ferromagnetic loss drum at its input mounted concentrically around a field system of alternate poles at its output. The drum is mechanically in-

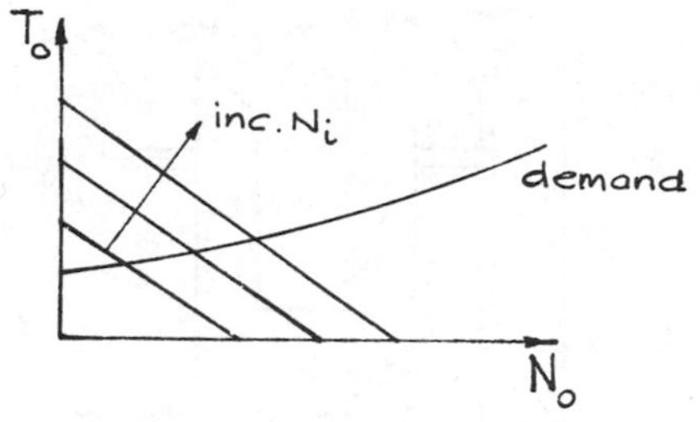

Fig. 3.2d. Matching a fluid coupling to a demand

dependent of the pole member, from which it is separated by a small radial air gap. When both members are stationary and the field system is excited, a magnetic field pattern of alternating polarity is established. If the field pattern is now rotated, eddy currents are induced in the drum, producing a torque which is a function of field current and slip speed. As for the fluid coupling we may again say that efficiency, η is given by

$$\eta = (1 - s)\, T_o/T_i,$$

where s is the slip, and again, under steady-state conditions, the torque ratio T_o/T_i is unity. Variable output speed is obtained simply by varying the slip, by controlling the excitation to the field system. The output speed can be varied from zero to the full speed of the drive, but, in practice, the speed range is limited by the heat-dissipation capacity of the drum. Since the eddy-current coupling is close in nature to the induction motor the torque-speed curve for a given input speed is almost identical to that of the induction motor (Fig. 2.1h), the torque being roughly proportional to the square of the field current. An eddy-current coupling is normally rated at about 50% of its maximum torque and is designed to deliver this torque at 4 - 8% slip.

In marine transmissions clutches are often used as switches to connect extra engines as required to the propeller via a common gearbox. For example in some modern naval vessels it is standard practice to employ a total of four turbines (two steam and two gas) to drive each of two propellers through a gearbox. The two steam turbines are used for cruising and the two gas turbines for high-speed operation. Clutches and couplings are also often used for simple gear changing operations (usually forward to reverse) as in cars. However when a slow speed diesel piston-engine is employed and direct-coupled to a propeller it is not uncommon to design the transmission with no clutch. In that case the starting device must be powerful enough to turn not only the engine but also the propeller. Such an arrangement is possible since a ship never has to climb a hill which would create a torque intercept on the demand characteristic. Also the propeller is itself a slipping element; the ship can be held at the quayside with the propeller turning. This is in contrast to a car which cannot be held with its wheels rotating.

For a gas-turbine engine the power turbine is usually separated from the compressor-turbine as shown in Fig. 3.2e. This arrangement has a torque characteristic which is then at least as favourable as the internal combustion

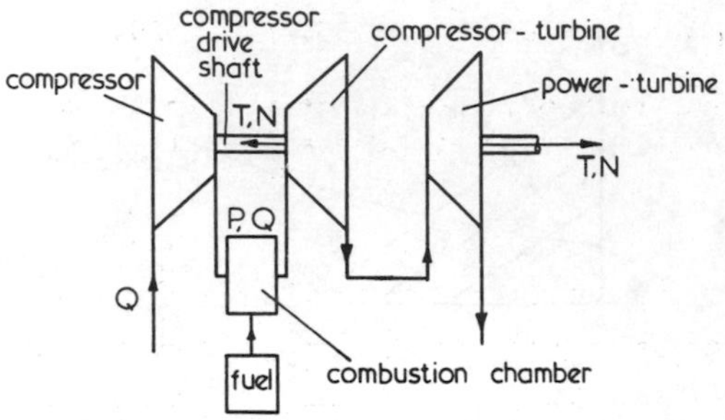

Fig. 3.2e. The compound gas-turbine engine

engine plus fluid-coupling for traction purposes. Increase in load causes a speed reduction of the power-turbine without alteration of the operating conditions of the compressor and the compressor-turbine (called collectively the gas generator). The latter can still deliver the maximum mass flow and thus enable a fairly constant output power to be obtained from the power turbine. Hence the output torque will increase with decrease in speed, and when the load is increased to reduce the power turbine speed to zero, it is possible to obtain a starting or 'stall' torque two to three times the torque delivered at full speed. Full advantage is thus taken of the form (ii) in Fig. 2.1a of the output characteristic of the power turbine.

Worked example. The results of experimental tests on a fluid coupling are given below, when the input shaft is rotated at 33.3 rev/s:

slip	1.0	0.5	0.2	0.1	0.06	0.04
torque Nm	87	62	39	27	19.5	15

(a) Plot a family of curves of torque carried versus input speed for different values of slip. Cover a range of input speed from 0-66.6 rev/s and a range of torque from zero to 100 Nm. Neglect Reynolds Number effects.

(b) The above coupling is fitted to a reciprocating engine which has the following full-throttle output characteristic:

speed rev/s	25.0	33.3	41.6	50.0	58.3	66.6
torque Nm	69	70	69	65	60	54

The load has a torque-speed characteristic given by

$$T = N \text{ Nm},$$

where N is in rev/s. What is the load speed, slip and engine speed under equilibrium conditions?

Solution. (a) From equation (3.2.1), $T = aN_i^2$ for a given slip. From the coupling characteristic given we have the following values of a:

slip	1.0	0.5	0.2	0.1	0.06	0.04
torque Nm	87	62	39	27	19.5	15
a Nm/(rev/s)^2 x 10^{-5}	79	56	35.4	24.5	17.6	13.6

Hence we may find the torque-slip characteristic at any other input speed. All such characteristics may be plotted as parallel lines on log-log graph paper (Fig. 3.2f).

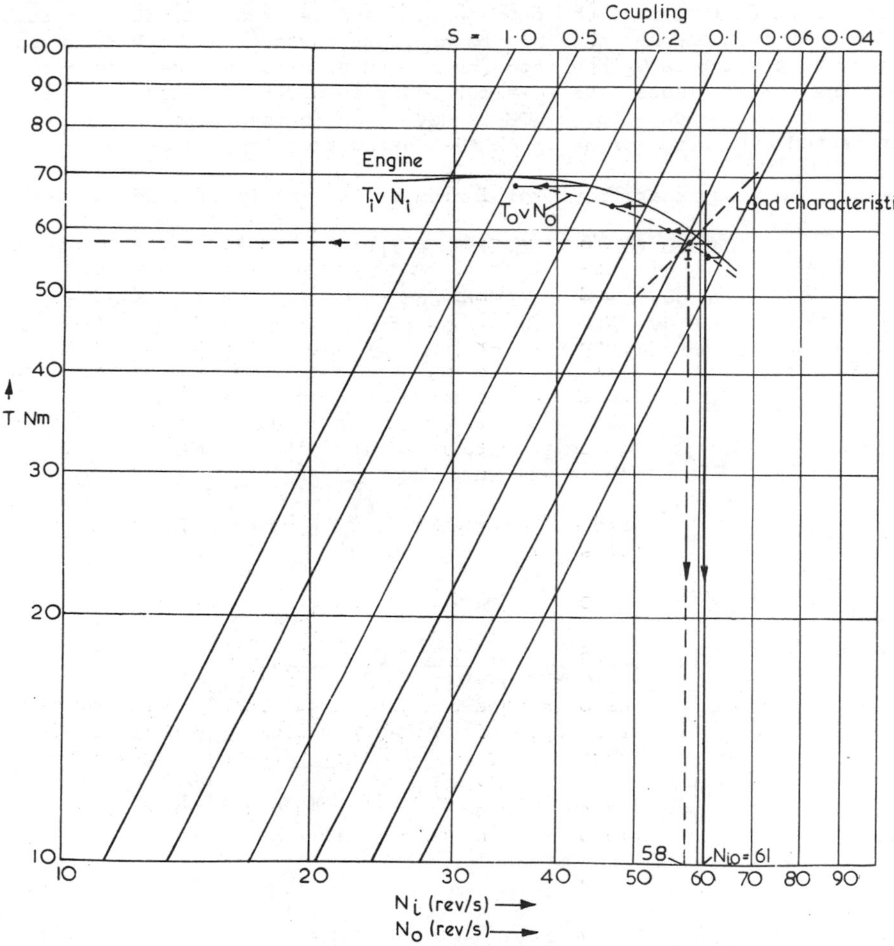

Fig. 3.2f. Matching an engine to a fluid coupling

(b) For each point of intersection with the engine characteristic we may read off the input (or engine) torques and the input (or engine) speeds N_i. Then, knowing s values, we can deduce the output speeds N_o. These are also the load speeds. Also the output torques T_o are equal to the input torques, and hence we may plot a graph of T_o versus N_o. Superimposing the load characteristic gives a point of intersection I defining the equilibrium values of output (load) torque and output (load) speed, from which the slip and the input (engine) speed N_{io} may also be deduced. The three values are 58 rev/s, 0.05 and 61 rev/s respectively.

3.3 The Torque Converter

As the name implies, the torque converter gives a 'conversion' of torque from input to output from one value to another (although 'amplifier' would be a better term). This is done by employing a stationary set of blades together with the usual pump and turbine rotors of the fluid coupling. These blades are fixed to the converter frame known as the stator and thus collect the working fluid as it leaves the pump portion and redirect it into the turbine portion. It then becomes possible to use curved vanes in the turbine so as to increase the torque T_o developed by it and so produce some torque conversion. The stator blades have some curvature in order to deliver the fluid into the turbine. Hence a torque T_{FL} is developed in the stator casing which is counteracted by fixing torque T_F on the casing from its foundation.

Hence we may say that, for equilibrium of externally imposed torques

$$\text{demand torque} \quad T_o = T_F + T_i. \tag{3.3.1}$$

Typical plots of torque ratio T_o/T_i versus N_o/N_i and of efficiency η versus N_o/N_i are shown in Fig. 3.3a. It may be seen that the torque ratio falls as the output speed increases from a 'stall' value of (typically) between 2 and 4

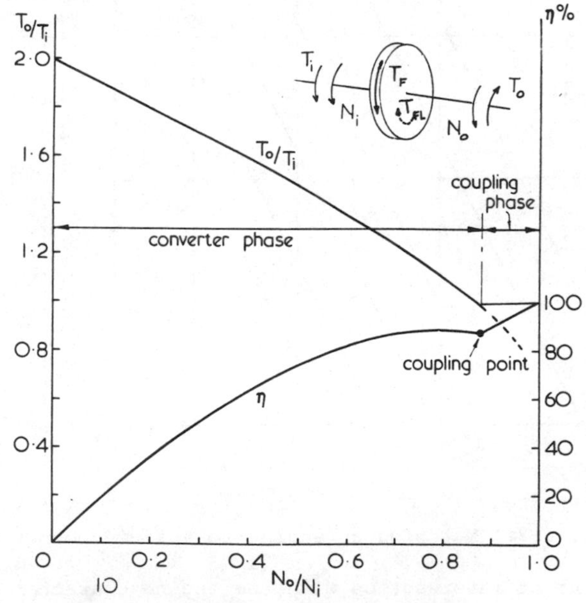

Fig. 3.3a. Two-phase converter characteristics

to about 0.84 at a speed ratio of about 0.95. It should be noted that the efficiency decreases for torque ratios less than about 1.1. Consider torque-ratios less than unity. In this region the efficiency varies from about 0.88

downwards, which is less than the efficiency of a fluid coupling operating at the same range of speed-ratio. In the latter the torque ratio is always unity, hence the efficiency T_oN_o/T_iN_i is equal to the speed ratio, in this case between about 0.88 and 1.0. It is obvious from equation (3.3.1) that as the output torque T_o becomes less than the input torque T_i so the fixing torque T_F from the stator changes sign. In fact, when T_o is less than T_i, then T_F acts in the same sense as T_o, that is the fluid acts on the stator in the same sense as T_i. If the stator is mounted on bearings and is provided with a free-wheel mechanism so that it is allowed to rotate in the same sense as the input shaft, but not in the opposite sense, the torque converter will operate normally when $T_o > T_i$ and as a fluid coupling for all speed ratios above that for which $T_o = T_i$, since then the stator can carry no torque. The changeover then occurs at what is known as the 'coupling point' (Fig. 3.3a). Such a converter is sometimes known as a two-phase converter or a converter-coupling. Torque converters are found in the fields of traction and earth moving. In automotive transmissions, for example, the torque converter has an advantage over the clutch and the fluid coupling in giving torque amplification (usually of between 2:1 and 1:1) for use during periods of vehicle acceleration. It thus combines the rôles of a slipping and a ratioing element.

Worked example. Fig. 3.3b shows the pressure-flow characteristics of the pump unit of a torque-converter for an input speed of 3000 rev/min together with the pressure-flow demand characteristics and the efficiency-flow characteristic of the turbine unit for different turbine output speeds 500-3000 rev/min.

Plot the torque-speed output characteristic of the torque converter for this input speed of 3000 rev/min to the pump unit. Assume that the pressure head generated by the pump unit is equal to that consumed in the turbine unit, that is no energy losses occur in the fixed blades.

Solution. Matching of the flow characteristics of the pump and turbine units must first be achieved. This is carried out by combining the pump and turbine unit curves of Fig. 3.3b as shown in Fig. 3.3c. At the points of intersection thus produced the output torque of the turbine unit may be deduced for different output speeds from a knowledge of its efficiency characteristic. To assist in this a table is constructed as follows, in which the turbine speeds, pressures and flows in Fig. 3.3c are converted to rad/s, N/m^2 and m^3/s respectively.

turbine O/P speed N_T rad/s	pressure P N/m^2	flow Q m^3/s	PQ Nm/s	turbine η_T	$\eta_T PQ$	turbine torque $= \frac{\eta_T PQ}{N_T}$ Nm
52	560,000	2.41×10^{-3}	1350	0.85	1140	22
105	520,000	3.08×10^{-3}	1600	0.85	1360	13
210	500,000	3.33×10^{-3}	1660	0.82	1360	6.5
315	470,000	3.58×10^{-3}	1680	0.78	1310	4.7

The torque speed output characteristic of the torque converter is then constructed and is shown in Fig. 3.3d.

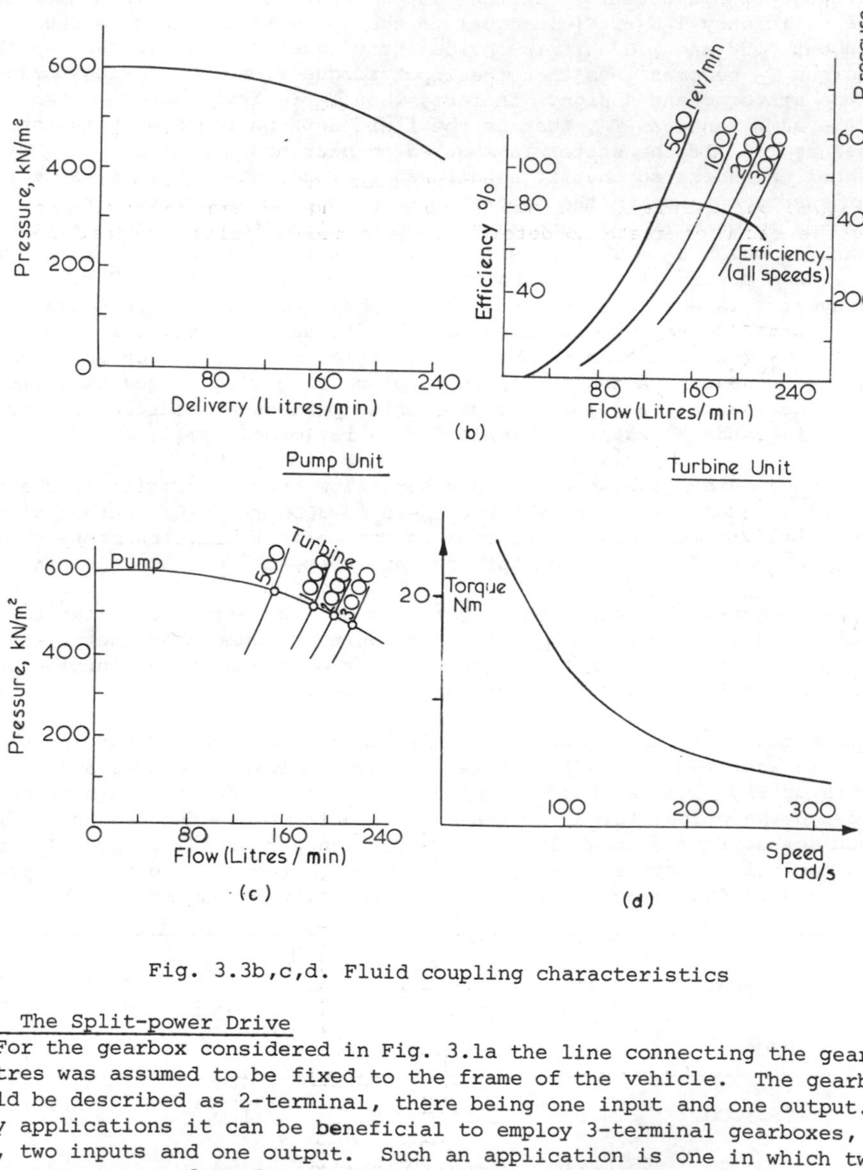

Fig. 3.3b,c,d. Fluid coupling characteristics

3.4 The Split-power Drive

For the gearbox considered in Fig. 3.1a the line connecting the gear-wheel centres was assumed to be fixed to the frame of the vehicle. The gearbox could be described as 2-terminal, there being one input and one output. In many applications it can be beneficial to employ 3-terminal gearboxes, having, say, two inputs and one output. Such an application is one in which two power converters are used to drive a single load. For example, in many marine drives, two turbines or diesel engines often drive a single propeller shaft via a three-terminal reduction gearbox. This constitutes a form of parallel operation where the output torques of each converter are summed but the speed of each must be the same. However, in some applications we may not wish to tolerate the limitation of an identical speed at each input or a set speed-ratio between outputs as in the simple gear train of Fig. 3.1d. A case in point is where a variable-speed device - a hydrostatic drive, say, cannot cope with the total output power of a power converter, perhaps owing to its fluid capacity being too small. With the aid of a different type of 3-terminal gearbox the variable-speed device can still be used, albeit to less ratioing effect.

3.5 The Epicyclic Gearbox

If we devise a 3-terminal gearbox by allowing the link connecting the gear-wheel centres to rotate as well as the gears themselves the resulting arrangement is known as a simple epicyclic gear. Such a gear is shown in Fig. 3.5a and its use in a transmission system in Fig. 3.5b. We can obtain the equation relating input speeds and output speed by referring all speeds to the input arm A (Fig. 3.5a). This is tantamount to reducing the arm speed to zero, thus allowing the ratio of teeth to determine the ratio of relative speeds.

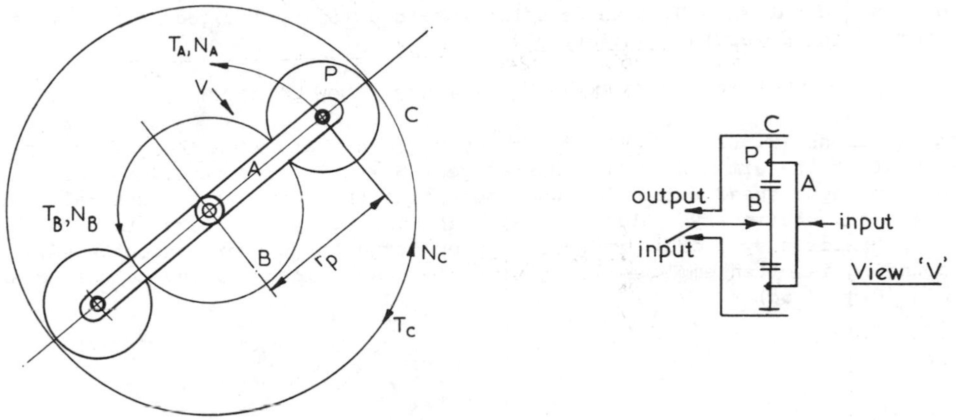

Fig. 3.5a. Simple epicyclic gear

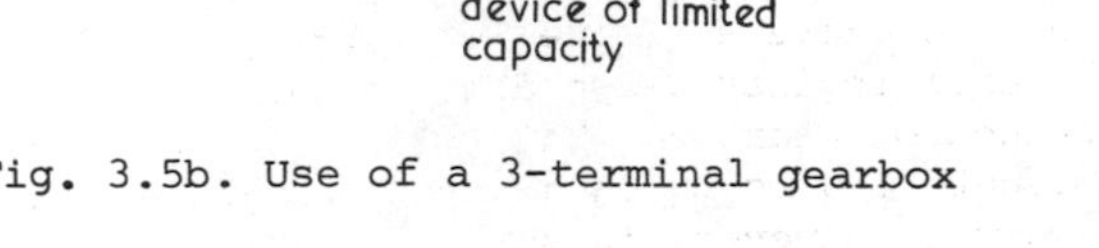

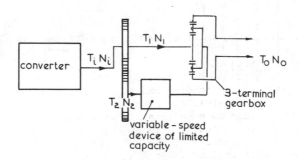

Fig. 3.5b. Use of a 3-terminal gearbox

Thus

$$(N_C - N_A)/(N_B - N_A) = (-1)B/C,$$

where N_A etc. denote speeds and B, C denote the numbers of teeth on gears B and C, respectively. The (-1) appears due to the fact that on bringing the arm A to rest, rotation N_C is opposite to rotation N_B.

Hence $$N_C = N_B\,(-B/C) + N_A\,(1 + B/C). \qquad (3.5.1)$$

It should be noted that the sum of the speed coefficients on the right hand side of equation (3.5.1) is unity. This is true whichever speeds are placed on the right-hand side, provided the coefficient of the speed on the left-hand side it itself unity. To obtain a simple gear train we may put

$$N_A = 0,$$

whence $$N_C = N_B\,(-B/C). \qquad (3.5.2)$$

If we instead put $N_B = 0$, then we still retain an epicyclic gear configuration, but now of the 2-terminal variety and

$$N_C = N_A\,(1 + B/C). \qquad (3.5.3)$$

From equations (3.5.2) and (3.5.3) it may be seen that, when the gear accepts the input at the arm A with the central gear B stationary, it gives a higher output to input speed ratio than when accepting the input at the central gear with the arm stationary. This is one reason why epicyclic gears are often used in preference to ordinary gears. When large torques are encountered, compounding is often employed so as to achieve some load sharing between gear-wheels (Fig. 3.5c).

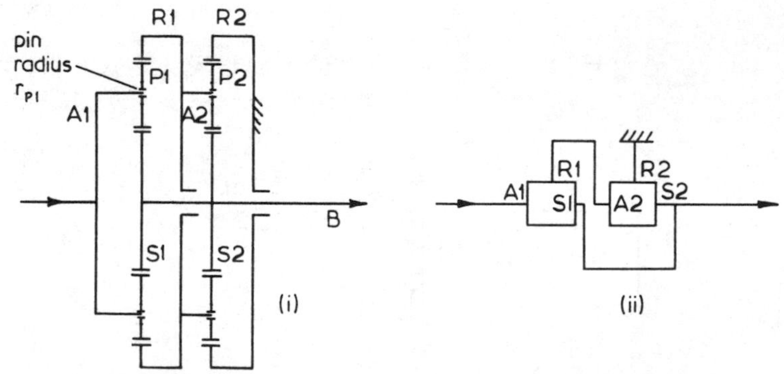

Fig. 3.5c. Compound epicyclic gear

Worked example. Deduce the speed ratios N_{R1}/N_{A1} and N_B/N_{A1} for the gear of Fig. 3.5c in which ring gear R_2 is fixed. Determine the velocity of rubbing of the planet gear P_1 on its pin of radius r_{p1}.

Solution

$$\frac{N_{S1} - N_{A1}}{N_{R1} - N_{A1}} = (-1)\ R1/S1,$$

$$\frac{N_{S2} - N_{A2}}{N_{R2} - N_{A2}} = (-1)\ R2/S2.$$

Note that in each equation reference is made to an arm (either A1 or A2).

Now $$N_{S1} = N_{S2} = N_B$$

$$N_{A2} = N_{R1}$$

and $$N_{R2} = 0$$

and hence $$\frac{N_B - N_{A1}}{N_{R1} - N_{A1}} = -\ R1/S1$$

$$\frac{N_B - N_{R1}}{-\ N_{R1}} = -\ R2/S2,$$

where R and S denote numbers of teeth. Hence it may readily be shown that

$$N_{R1}/N_{A1} = (R1 + S1)S2/|S1(R2 + S2) + R1S2|$$

and $$N_B/N_{A1} = (R1 + S1)(R2 + S2)/|S1(R2 + S2) + R1S2|.$$

Alternatively we may consider each cell of the compound epicyclic, using an equation similar to equation (3.5.1). The cells may be represented as shown in Fig. 3.5c(ii) in which A represents an arm, S a sun gear and R a ring gear.

Thus $$N_{R1} = -\ N_{S1}\frac{S1}{R1} + N_{A1}\ (1 + \frac{S1}{R1})$$

$$N_{R2} = -\ N_{S2}\frac{S2}{R2} + N_{A2}\ (1 + \frac{S2}{R2})$$

in which $$N_{R1} = N_{A2}$$

$$N_{S1} = N_{S2} = N_B$$

$$N_{R2} = 0.$$

These give the same results as before for the speed ratios N_{R1}/N_{A1} and N_B/N_{A1}.

Finally, rubbing velocity, v_r on the pin of planet P1 = $(N_{P1} - N_{A1})r_{p1}$.

Now $$\frac{N_{p1} - N_{A1}}{N_{s1} - N_{A1}} = -\ S_1/P_1.$$

Hence
$$N_{p1} - N_{A1} = (S_1/P_1)(N_{A1} - N_B).$$

Thus
$$v_r = (S_1/P_1)(N_{A1} - N_B)\, r_{p1}.$$

A very common use of an epicyclic gearbox is in automotive transmissions which embody a fluid coupling or a torque converter rather than a friction clutch as the slipping element. Whereas the input and output plates of a friction clutch can be separated quite cleanly for gear changing (Fig. 3.2a), this is not true of a fluid coupling or of a torque converter and a complete disconnection of gearbox from engine is thus impossible. A way round this problem is to utilise 3-terminal or epicyclic gears in the gearbox where one terminal of a particular epicyclic train may be brought to rest at will, depending upon the gear ratio to be selected. Such a gearbox, known as the Wilson gearbox is shown diagrammatically in Fig. 3.5d. This can cater for four speed-ratios and reverse. Top

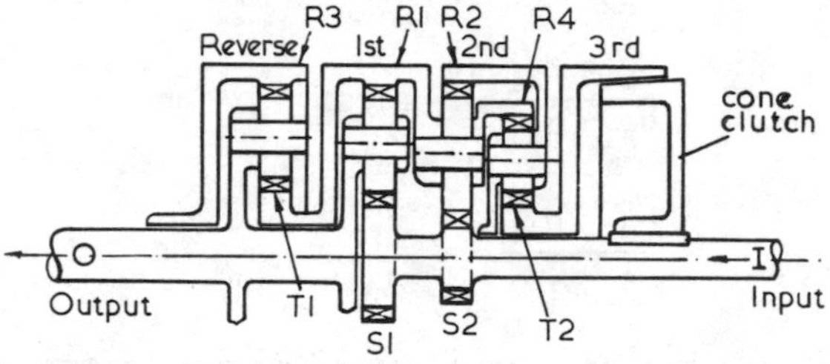

Fig. 3.5d. The Wilson gearbox

gear is obtained by engaging the cone-clutch, which is in effect a one to one gear. First, second, third and reverse are obtained by applying brake bands in turn to the drums as indicated, the cone-clutch being disengaged.

Consider the general case applicable to the selection of any gear-ratio.

Thus
$$\frac{N_{S2} - N_{T1}}{N_{R2} - N_{T1}} = -\frac{R2}{S2},$$

the relevant arm being attached to gear T1.

Also
$$\frac{N_{S1} - N_o}{N_{R1} - N_o} = -\frac{R1}{S1},$$

the relevant arm being attached to output shaft O.

Further
$$\frac{N_{T2} - N_{R2}}{N_{R4} - N_{R2}} = -\frac{R4}{T2},$$

the relevant arm being attached to gear R2,

and
$$\frac{N_{T1} - N_o}{N_{R3} - N_o} = -\frac{R3}{T1},$$

the relevant arm being attached to output shaft O.

Now $$N_{S1} = N_{S2} = N_I$$

and $$N_{T1} = N_{R4} = N_{R1}.$$

Thus the above equations may be rewritten

$$\frac{N_I/N_o - N_{R1}/N_o}{N_{R2}/N_o - N_{R1}/N_o} = -\frac{R2}{S2},$$

$$\frac{N_I/N_o - 1}{N_{R1}/N_o - 1} = -\frac{R1}{S1},$$

$$\frac{N_{T2}/N_o - N_{R2}/N_o}{N_{R1}/N_o - N_{R2}/N_o} = -\frac{R4}{T2}$$

and $$\frac{N_{R1}/N_o - 1}{N_{R3}/N_o - 1} = -\frac{R3}{T1}.$$

The speed ratios featured in these four equations are

$$N_I/N_o,\ N_{R1}/N_o,\ N_{R2}/N_o,\ N_{T2}/N_o \text{ and } N_{R3}/N_o.$$

For 1st, 2nd, 3rd and reverse gears, one of these ratios will be set to zero by the appropriate band brake, depending upon the gear selected, leaving four ratios to be found from the four equations. For top gear, by virtue of connecting the cone-clutch,

$$N_{T2} = N_I$$

and hence

$$\frac{N_{T2}}{N_o} = \frac{N_I}{N_o},$$

again leaving four ratios to be found from the four equations.

A class of epicyclic gear deserving special mention is the bevel epicyclic gear which, for example, finds use in the so-called differential gear. Consider again equation (3.5.1) in which we now require that N_A be the mean of N_B and N_C. This in turn means that B/C must equal unity, that is the number of teeth on B must be equal to the number of teeth on C. Clearly an epicyclic gear of the form shown in Fig. 3.5a (View V) is ruled out. How then can we redesign this epicyclic gear to give the desired tooth ratio? The answer is to use bevel planets in the epicyclic train as shown in Fig. 3.5e. It then becomes possible to make the mean diameters of gear wheels B and C equal and hence the numbers of teeth equal. The prime example of the use of such a 3-terminal bevel epicyclic gear is in the transmission of automobiles. The engine power is transmitted via the gearbox to the arm A and, when the car is travelling along a straight path, the planets do not spin on the arm A but simply push the gears B and C around at equal speeds. These gears are integral with the driving wheel axles and hence the driving wheels rotate at equal speeds. If, however, the automobile turns a corner, this means that one driving wheel, say that connected to gear C, is forced by the resistive torque from the ground to rotate more slowly than the other driving wheel. By virtue of the differential gear the latter driving wheel is automatically made to rotate faster than before, in accordance with the equation

$$N_B = 2N_A - N_C.$$

This allows the vehicle to turn the corner with adequately small scuffing between the road wheels and the ground.

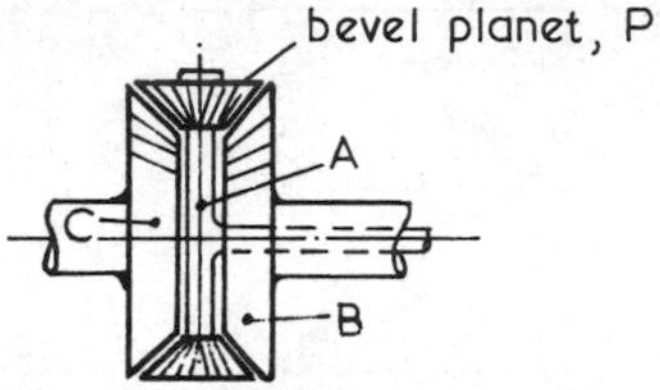

Fig. 3.5e. Bevel epicyclic gear

The earliest known use of a differential gear was by the ancient Chinese about 2400 B.C. The gear was incorporated in the so-called South Pointing Chariot which was supposedly used to cross the Gobi desert. This is referred to in a question at the end of this chapter.

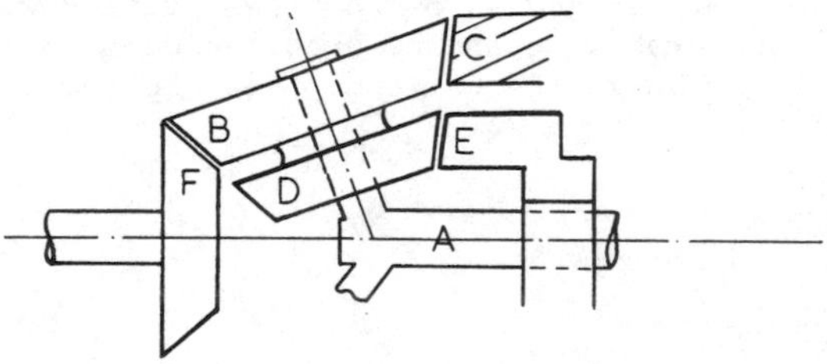

Fig. 3.5f. Humpage's gear

The 4-terminal epicyclic gear. Fig. 3.5f shows an epicyclic gear known as Humpage's Gear used in a four-terminal rôle. Thus,

$$\frac{N_F - N_A}{N_C - N_A} = -\frac{C}{F}$$

and

$$\frac{N_F - N_A}{N_E - N_A} = -\frac{B}{F} \cdot \frac{E}{D}.$$

These two equations feature four unknown speeds and to solve them we must specify independent speeds at two of the terminals, from which we shall obtain the interelated speeds at the other two terminals. It should be noted that we cannot specify independent speeds at *three* terminals since this leaves only one unknown to be found from two equations, neither of which is redundant. For example suppose we are told that gear-wheel C is held stationary and we are asked to deduce the speeds N_E and N_F in terms of a given speed N_A.

Thus
$$N_C = 0.$$

Hence
$$N_F/N_A - 1 = \frac{C}{F}$$

and
$$\frac{N_F/N_A - 1}{N_E/N_A - 1} = \frac{EB}{FD}\,(-1).$$

Thus
$$N_F/N_A = 1 + \frac{C}{F}$$

and
$$N_E/N_A = 1 - \frac{CD}{EB}.$$

The last equation shows that by a suitable choice of teeth on the compound planet BD, both the speed and the direction of rotation of E relative to A may be prescribed.

3.6 Torque Relations in a Simple Epicyclic Gear

Considering the *external* torques applied to an epicyclic gear, the following rules may be used to considerable advantage:

(1) For an input the external (or propulsive) torque is in the same direction as the rotation.
(2) For an output the external (or resistive) torque is in the opposite direction to the rotation.
(3) For no acceleration the external torques must balance.
(4) For no losses the input power and the output power must balance.

Applying these rules in Fig. 3.5a gives

$$T_A + T_B = T_C$$

$$T_A N_A + T_B N_B = T_C N_C, \qquad (3.6.1)$$

whence
$$T_A = T_C(N_C - N_B)/(N_A - N_B)$$

and
$$T_B = T_C(N_C - N_A)/(N_B - N_A)$$

Using equation (3.5.1), namely

$$N_C = N_B(-B/C) + N_A\,(1 + B/C),$$

then
$$T_A = T_C.(1 + B/C)$$

and
$$T_B = T_C\,(- B/C) \qquad (3.6.2)$$

Losses do of course occur in gears, but if the gears are well designed and lubricated such losses at a meshing pair amount to only about 0.5% of the power transmitted through that pair. Equation (3.6.1) may then be slightly modified to account for these losses.

Worked example. Deduce for the compound epicyclic gear of Fig. 3.5c the torque ratio T_{S1}/T_{A1} and the power ratio $T_{S1}N_{S1}/T_{A1}N_{A1}$.

Solution We may treat the first section of gearing as having an input at A1, with outputs at R1 and S1. Also, in order to preserve the speed ratios N_B/N_{A1} and N_{R1}/N_{A1} which were deduced in section 3.5, we must assume all speeds to be in the same direction, say clockwise looking from the right. Hence

$$T_{A1} N_{A1} = T_{R1} N_{R1} + T_{S1} N_{S1}$$

or

$$1 = \frac{T_{R1}}{T_{A1}} \cdot \frac{N_{R1}}{N_{A1}} + \frac{T_{S1}}{T_{A1}} \cdot \frac{N_B}{N_{A1}}. \tag{3.6.3}$$

Also, since T_{A1} is an input torque (and therefore in the same direction as N_{A1}) and since T_{R1} and T_{S1} are output torques (in the opposite directions to N_{R1}, and N_{S1} respectively), then

$$T_{A1} = T_{R1} + T_{S1}$$

or

$$1 = \frac{T_{R1}}{T_{AI}} + \frac{T_{S1}}{T_{A1}}. \tag{3.6.4}$$

In a similar fashion we may treat the second section of gearing as having inputs at S1 and R1 and an output at S2, but not forgetting that a fixing torque T_{R2} will exist on ring-gear R2.

Thus

$$T_{R1} N_{R1} + T_{S1} N_{S1} = T_B N_B$$

or

$$\frac{T_{R1}}{T_{A1}} \cdot \frac{N_{R1}}{N_{A1}} + \frac{T_{S1}}{T_{A1}} \cdot \frac{N_B}{N_{A1}} = \frac{T_B}{T_{A1}} \cdot \frac{N_B}{N_{A1}}. \tag{3.6.5}$$

Also, assuming fixing torque T_{R2} is anti-clockwise, say,

$$T_{R1} + T_{S1} = T_B + T_{R2}.$$

Thus

$$\frac{T_{R1}}{T_{A1}} + \frac{T_{S1}}{T_{A1}} = \frac{T_B}{T_{A1}} + \frac{T_{R2}}{T_{A1}}. \tag{3.6.6}$$

The four equations (3.6.3) to (3.6.6) feature the four unknown torque ratios

$$\frac{T_{R1}}{T_{A1}}, \frac{T_{S1}}{T_{A1}}, \frac{T_B}{T_{A1}} \text{ and } \frac{T_{R2}}{T_{A1}}.$$

Hence these may be found. The value of T_{S1}/T_{A1} in particular is given by

$$\frac{T_{S1}}{T_{A1}} = \frac{1 - \frac{N_{R1}}{N_{A1}}}{\frac{N_B}{N_{A1}} - \frac{N_{I1}}{N_{A1}}}.$$

Using the speed ratios N_{R1}/N_{A1} and N_B/N_{A1} found in section 3.5 gives

$$T_{S1}/T_{A1} = S_1/(R1 + S1). \tag{3.6.7}$$

Finally this gives the required power-ratio,

$$\frac{T_{S1}}{T_{A1}} \cdot \frac{N_{S1}}{N_{A1}} = \frac{S1(R2 + S2)}{S1(R2 + S2) + R1S2} .$$

Alternatively we may deduce the torque ratio T_{S1}/T_{A1} by again considering the compound gear as consisting of two cells and using equations (3.6.2). Thus, with the notation of Fig. 3.5c (ii) we have

$$T_{A1} = T_{R1}\ (1 + \frac{S1}{R1})$$

$$T_{S1} = T_{R1}\ (- \frac{S1}{R1})$$

$$T_{A2} = T_{R2}\ (1 + \frac{S2}{R2})$$

and

$$T_{S2} = T_{R2}\ (- \frac{S2}{R2}) .$$

Hence

$$\frac{T_{S1}}{T_{A1}} = \frac{-\frac{S1}{R1}}{1+\frac{S1}{R1}} = \frac{- S1}{R1 + S1} ,$$

in which T_{S1} is a demand torque on the first cell rather than an input torque to the second cell. Hence the negative sign. This result agrees with that obtained previously (equation (3.6.7)).

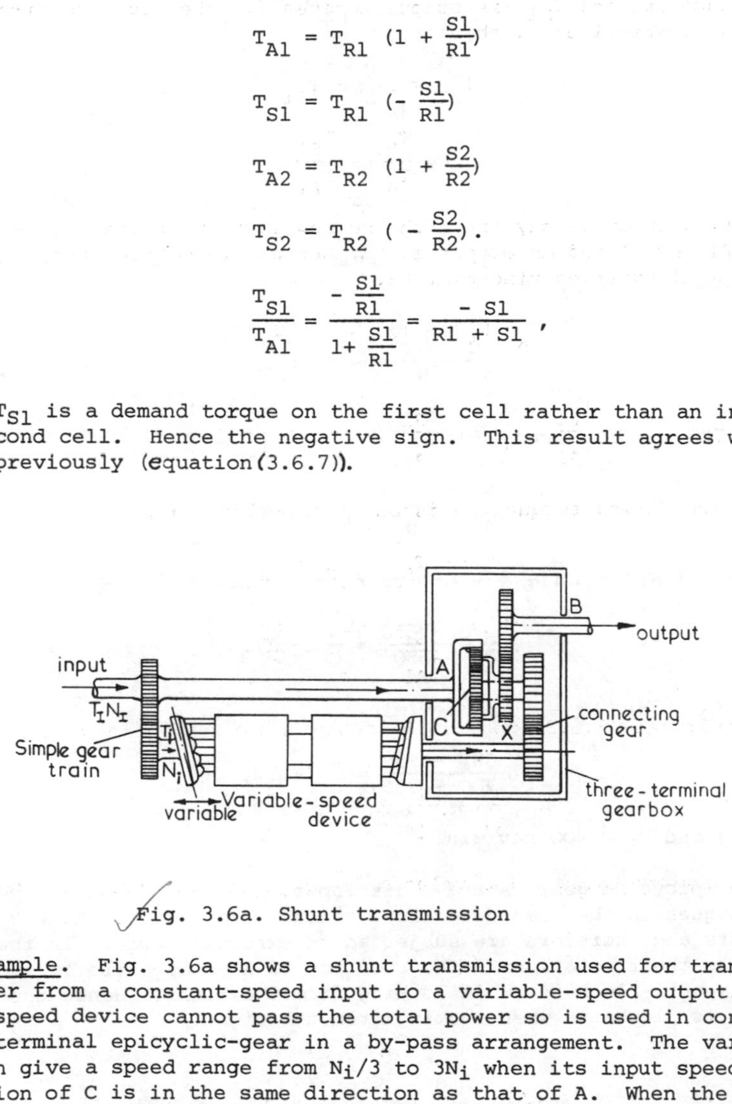

Fig. 3.6a. Shunt transmission

<u>Worked example</u>. Fig. 3.6a shows a shunt transmission used for transmitting a large power from a constant-speed input to a variable-speed output. The variable-speed device cannot pass the total power so is used in conjunction with a 3-terminal epicyclic-gear in a by-pass arrangement. The variable-speed device can give a speed range from $N_i/3$ to $3N_i$ when its input speed is N_i and the rotation of C is in the same direction as that of A. When the output speeds N_c from the variable-speed device are $N_i/3$ and $3N_i$ rev/min, the output

speeds from B are to be 650 and 950 rev/min respectively in the opposite direction to the 700 rev/min input at A.

(a) If the ratio (teeth on B/teeth on X) is unity deduce the ratio (teeth on A/teeth on C) and speed N_i.

(b) Assuming no losses, what is the torque carried by the variable-speed device when the torque delivered at B is 500 Nm at 950 rev/min?

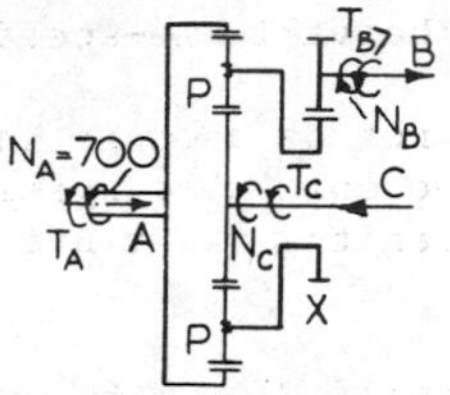

Fig. 3.6b. 3-terminal gearbox

Solution. (a) Referring to Fig. 3.6b, which is a diagrammatic representation of the 3-terminal epicyclic gear shown in Fig. 3.6a,

$$\frac{N_C - N_X}{N_A - N_X} = \frac{A}{C}(-1).$$

Now N_A = + 700 rev/min and hence

$$\frac{N_C - N_X}{700 - N_X} = \frac{A}{C}(-1).$$

But when $N_C = + N_i/3$ then N_B = - 650 rev/min and hence N_X = + 650 rev/min.

Thus $$\frac{N_i/3 - 650}{700 - 650} = \frac{A}{C}(-1).$$

Also, when $N_C = 3N_i$, N_X = + 950 rev/min.

Thus $$\frac{3N_i - 950}{700 - 950} = \frac{A}{C}(-1),$$

whence $\frac{A}{C} = 7$ and N_i = 900 rev/min.

(b) For the epicyclic gear, A and C are inputs and therefore are subjected to external torques in the same directions as their rotations, while B and hence X are outputs and therefore are subjected to external torques in the opposite directions to their rotations. The rotation of X is opposite to that of B and hence is anti-clockwise when looking from the right. Hence the external torque on X is clockwise when looking from the right.

Thus $$T_A + T_C = T_X$$

and $$T_A N_A + T_C N_C = T_X N_X.$$

Since the teeth on B and X are equal, T_X = 500 Nm.

Also $$N_A = +\ 700 \text{ rev/min},\ N_X = +\ 950 \text{ rev/min}$$

and $$N_C = 3 \times 900 = 2700 \text{ rev/min}.$$

Hence $$T_C = +\ 63.5 \text{ Nm}.$$

This is the torque carried by the variable-speed device.

It should be noted that T_B may not be used in the epicyclic torque equation since shaft B is not part of the epicyclic system. In general there will be a torque fixing the frame between the shaft B and shaft X, although in this case it is zero since $T_B = T_X$.

Let us now look at the system of Fig. 3.6a in a more general way. We have already seen by comparison with equation (3.5.1) that, for the epicyclic gear shown in Figs. 3.6a, b,

$$N_C = aN_A + xN_X \tag{3.6.8}$$

where $$a + x = 1.$$

The overall speed ratio R is thus given by

$$R = N_X/N_A = (N_C/N_A - a)/(1 - a).$$

Now $$N_C/N_A = (N_C/N_i) \cdot (N_i/N_A) = R_v G,$$

where $R_v = N_C/N_i$ = speed ratio through the variable speed device plus connecting gear

and $G = N_i/N_A$ = speed ratio through the simple gear train.

Hence $$R = (GR_v - a)/(1 - a), \tag{3.6.9}$$

Consider now the power carried by the variable-speed device as a proportion of the total output power transmitted. This proportion $\overline{P}$ may be termed the output-power ratio and be written,

$$\overline{P} = T_C N_C / T_X N_X. \tag{3.6.10}$$

But for torque balance in the epicyclic gear

$$T_A + T_C = T_X, \tag{3.6.11}$$

and for no losses in the simple gear-train,

$$T_I\ N_I = T_i\ N_i + T_A\ N_A.$$

For no losses in the variable-speed device or in the complete system, we may write

$$T_I\ N_I = T_X\ N_X = T_C\ N_C + T_A\ N_A$$

or $$T_A = (T_X N_X - T_C N_C)/N_A .$$

Using equation (3.6.11) gives

$$T_X N_X - T_C N_C + T_C N_A = T_X N_A$$

or $$T_X (N_X - N_A) = T_C (N_C - N_A).$$

Substitution into equation (3.6.10) then gives

$$\overline{P} = \frac{N_X - N_A}{N_C - N_A} \cdot \frac{N_C}{N_X} = \frac{1 - 1/R}{1 - 1/GR_v} .$$

Using equation (3.6.9) this then gives

$$\overline{P} = 1/(1 - a/GR_v). \tag{3.6.12}$$

It should be remembered that all speeds have been assumed to be in the same direction and so G and R_V may well take negative values in some circumstances, as indeed they do in Fig. 3.6a.

In some instances a shunt drive is used when the versatility of a variable-speed device is required, but where it may be too inefficient to be used on its own. Such inefficiency may arise from losses occurring in the energy conversions taking place within a variable-speed device. A mechanical path, on the other hand, does not suffer from such considerable losses. We can account for the losses in the variable-speed device by writing the power through C,

$$P_C = T_C N_C = \eta_v T_i N_i = \eta_v P_i, \text{ say}$$

where η_V is the efficiency of the variable-speed device. The overall efficiency η is then given by

$$\eta = P_X/(P_A + P_i) = P_X/(P_X - P_C + P_i)$$

$$= \frac{P_X/P_C}{(P_X/P_C) - 1 + (P_i/P_C)} .$$

Now from equation (3.6.10)

$$P_X/P_C = 1/\overline{P}.$$

Thus $$\eta = \frac{1/\overline{P}}{(1/\overline{P}) - 1 + (1/\eta_V)} = \frac{1}{1 - \overline{P}(1 - 1/\eta_v)} . \tag{3.6.13}$$

The particular arrangement of Fig. 3.6a obtained when the variable-speed device is a fluid-coupling is known as a torque-divider. The whole arrangement can then be drawn in a more compact manner as shown in Fig. 3.6c, from which it may be observed that

$$G = N_i/N_A = 1$$

and $$R_V = N_C/N_i = 1 - S_c,$$

where S_c is the coupling slip.

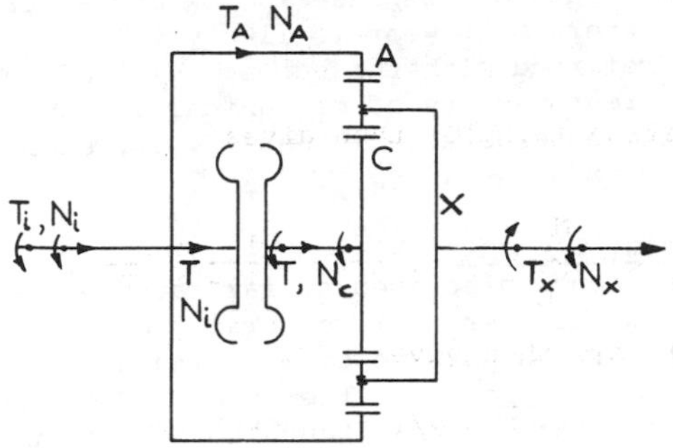

Fig. 3.6c. The torque divider

Also, referring to equation (3.6.8)

$$a = - A/C.$$

Hence using equation (3.6.9), the overall speed ratio R is given by

$$R = N_x/N_A = (GR_V - a)/(1 - a) = \frac{(1 - S_C) + A/C}{1 + A/C}.$$

Since no torque is 'lost' through the torque-divider the above ratio also represents its efficiency in terms of output power divided input power.

Also from equation (3.6.12), the output-power ratio

$$\bar{P} = (1 - \frac{a}{GR_V})^{-1} = \frac{1 - S_C}{(1 - S_C) + A/C}.$$

Thus, as in the case of Fig. 3.6a, the use of a 3-terminal epicyclic gear has resulted in a reduction of power to be passed by the variable speed device (the fluid coupling), while still utilising its slip capability.

An interesting variant of the shunt transmission is a transmission using mechanical and electrical elements rather than mechanical and fluid. An example of such a transmission is the so-called Thomas Drive which embodies an electric generator and an electric motor in place of the hydraulic pump and motor of the fluid-mechanical drive. Variations in speed are effected by varying the voltage and current to the motor.

3.7 Further Ratio Elements

It has been seen that the output speed of a ratio element such as a gearbox, say, is a predetermined ratio of the input speed. Belts, although having a similar function, are less predictable owing to slipping, stretching, or breaking due to excessive torque transmission. However, they constitute a convenient arrangement for transmitting power quietly over large distances. Different types of belt arrangement may be used for the same shaft dispositions which are found utilising gearing, namely parallel, intersecting and skew and the speed ratio is decided by the ratio of pulley diameters. Belts find an interesting use in some types of manual automotive transmissions in which long coned pulleys are employed at input and output. By this means the speed-ratio may be made continuously variable as the belt is moved axially along the pulleys. However, the speed of response is low and accuracy of ratio is limited. In applications where belt wear is likely to be excessive chains and sprockets are often preferred with the accompanying advantages of no slip or stretch. There is a great variety of mechanical variable-speed drives on the market, some exhibiting considerable ingenuity, for example the Kopp, the Floyd and the Davall-Metron variators.

3.8 Function-generating Elements

The transmission elements discussed so far have been either ratioing or slip devices. There is, however, a large branch of transmission concerned with mechanisms which for the purpose of matching a converter to a load, provide an output which is a function of the input. One example is the reciprocating mechanism which transmits the power from the piston to the crankshaft in an automotive engine (Fig. 1.2a). The need for this particular mechanism arises from the fact that power is most amenable to subsequent conversion when associated with rotational speed rather than with reciprocation. Other examples occur when a reciprocation or rocking motion is required in specialised equipment whose input is in the form of a constant torque at constant speed as, for example, supplied by an electric motor.

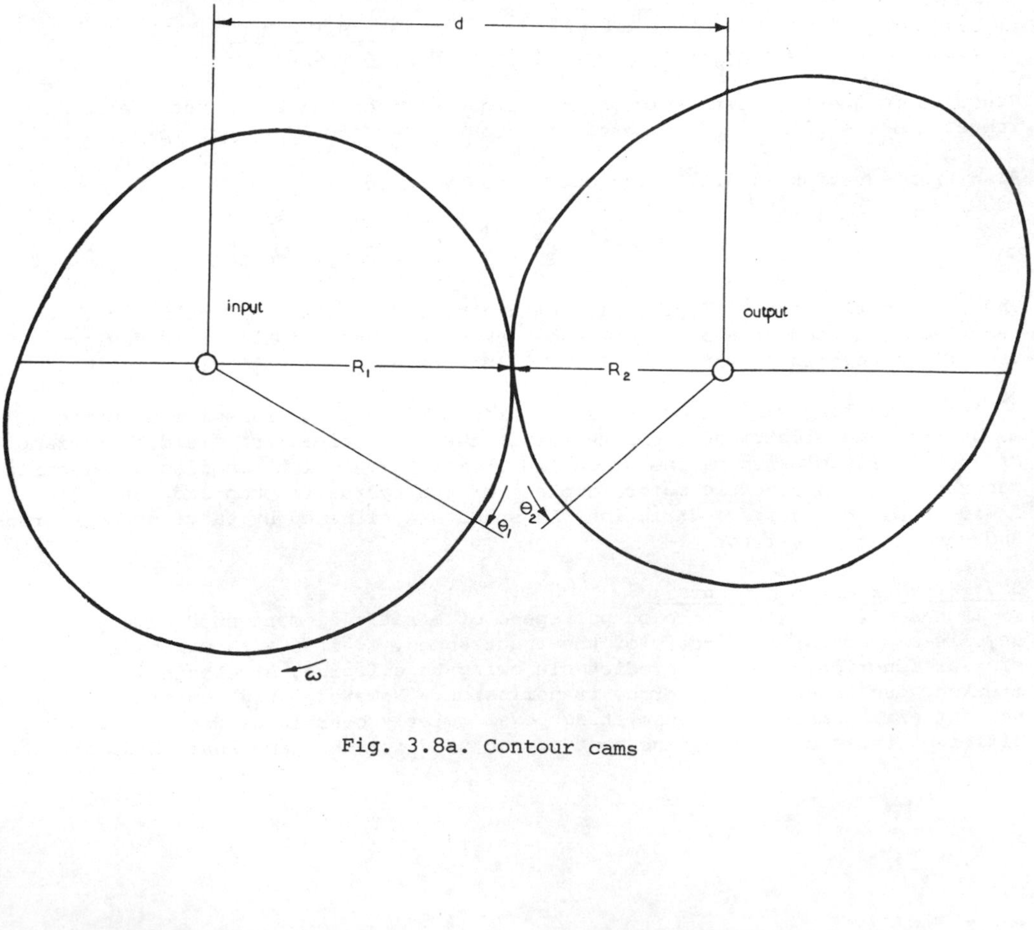

Fig. 3.8a. Contour cams

With the condition that the input must be at constant speed let us consider sample forms that a function-generating element can take (Figs. 3.8a to d). These forms depend on whether continuous rotation or a rocking motion is required at the output. Figs. 3.8a and b show two ways of producing continuous rotation while Figs. 3.8c and d show ways of producing rocking motion. Where-ever cams (Figs. 3.8a and c) are used the precise required output motion can be obtained, but linkages can be used more cheaply when a precise form is not important. Let us consider briefly the kinematic design of both cams and linkages.

Cams which feature continuous rotation at the output are known as contour cams (or non-circular gears) (Fig. 3.8a) and their design may be conducted on the following basis:

For no slipping $$R_1 \dot{\theta}_1 = R_2 \dot{\theta}_2,$$

where $$\dot{\theta}_1 = \omega = \text{constant}.$$

Thus $$R_1\omega = R_2 \dot{\theta}_2.$$

Also, for continuous contact with no slipping,

$$R_1 + R_2 = d.$$

Thus $$R_2 = d/(\dot{\theta}_2/\omega + 1).$$

Now one cam must complete a whole number of revolutions in the time during which the other cam completes one revolution. For a one to one ratio of revolutions, for example,

$$\int_{t=0}^{t=2\pi/\omega} d\theta_2 = 2\pi$$

or $$\int_0^{2\pi/\omega} \dot{\theta}_2 \, dt = 2\pi.$$

Also R_2 must be periodic in 2π intervals of θ_2 or in $2\pi/\omega$ intervals of t. For example, if we wish to design a pair of contour cams such that

$$\dot{\theta}_2 = \omega(1 + 0.5 \cos \omega t)$$

say, then we can first verify that

$$\int_0^{2\pi/\omega} \dot{\theta}_2 \, dt = 2\pi,$$

then $$R_2 = d/(2 + 0.5 \cos \omega t),$$

which is seen to be periodic in $\frac{2\pi}{\omega}$ intervals of t.

It now remains to establish the relationships R_1 (θ_1) and R_2 (θ_2).

Since
$$R_2 = d/(2 + 0.5 \cos \omega t),$$
then
$$R_1 = d - R_2 = d|1 - 1/(2 + 0.5 \cos \omega t)|$$
$$= d.\frac{1 + 0.5 \cos \omega t}{2 + 0.5 \cos \omega t},$$
in which $\omega t = \theta_1$.

Thus
$$R_1\ (\theta_1) = d\ .\ \frac{1 + 0.5 \cos \theta_1}{2 + 0.5 \cos \theta_1}\ .$$
Now
$$\theta_2 = \int \dot{\theta}_2\ dt = \omega \int (1 + 0.5 \cos \omega t)\ dt$$
$$= \omega t + 0.5 \sin \omega t$$
$$= \theta_1 + 0.5 \sin \theta_1.$$

Hence, for each value of θ_1 we may obtain R_1, θ_2 and R_2, and so the contour cams may be constructed. This particular pair is illustrated in Fig. 3.8a.

Had we been asked to design a pair of contour cams to generate
$$\dot{\theta}_2 = \omega\ (1 + \cos \omega t),$$
then we would have found that
$$R_1 = d\ .\ \frac{1 + \cos \omega t}{2 + \cos \omega t}.$$

This means that at $t = \pi/\omega, R_1 = 0$, which is a situation impossible to achieve in real hardware. The way round this problem is to avoid a zero value in $\dot{\theta}_2$ by designing for
$$\dot{\theta}_2 = \omega\ (1 + k + \cos \omega t),\ \text{say}$$
and including a differential in the system which will later subtract k.

Then we would have made
$$R_1 = d\ (1 + k + \cos \omega t)/(2 + k + \cos \omega t)$$
$$R_2 = d/(2 + k + \cos \omega t)$$
and the complete system would have been as shown in Fig. 3.8e.

If an output shaft is required to exhibit two speed oscillations during one revolution a pair of elliptical cams rotating about their centres may be used.

Cams with a reciprocating follower (Fig. 3.8c) are rather easier to construct. We may simply say that
$$x - x_o = R - R_o \text{ or } R = x - x_o + R_o = x$$
where R_o is some datum value of R at which $x = x_o$. Hence, knowing the required values of x(t) we may construct R(θ) on the assumption that the cam rotates with constant speed ω. If required a rack and pinion may be used to provide

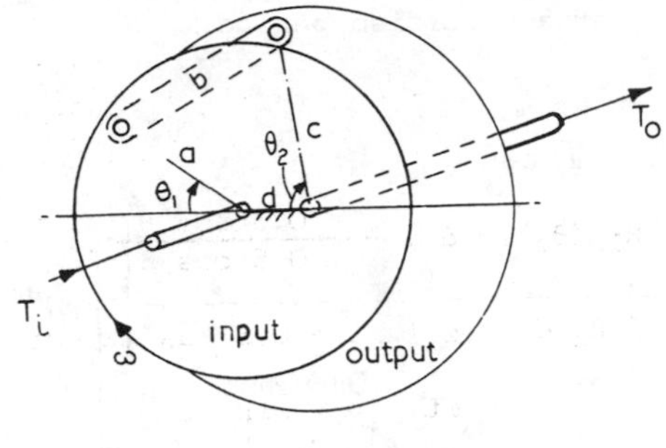

Fig. 3.8b. Drag-link mechanism

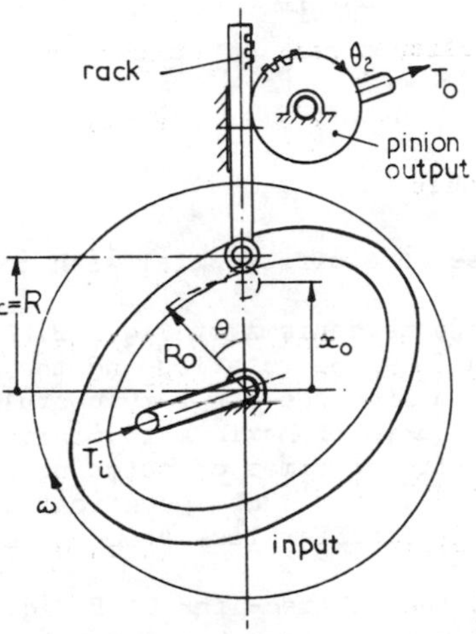

Fig. 3.8c. Cam and reciprocating follower

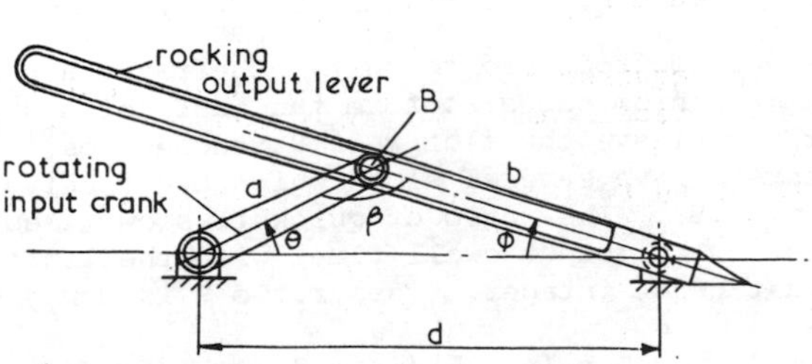

Fig. 3.8d. Quick-return mechanism

ultimate angular reciprocation. Cams are particularly used where the required output is given as a series of numerical values, a typical case being where a dwell at the output is required as often happens when say a valve is required to lift and remain open for a specified period.

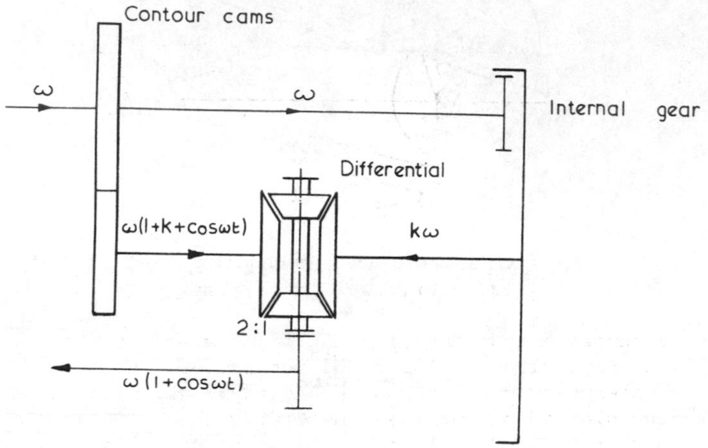

Fig. 3.8e. Use of a differential with contour cams

Consider now the linkage mechanisms of Figs. 3.8b and 3.8d. The former produces continuous variable-speed rotation and the latter an angular rocking motion at its output. Such linkages are very useful if subject to a broad specification such as the ratio of maximum to minimum output angular velocities (Fig. 3.8b) or the ratio of times of outgoing swing to return swing of the output lever (Fig. 3.8d). For such specifications they are cheaper to manufacture than are cams and suffer less from the effects of wear.

Consider the motion of the slotted-link b of Fig. 3.8d for rotation of crank a at constant angular velocity. For a maximum excursion of link b, $\beta = \pi/2$ and hence $\cos\theta = a/d$. This gives as the time ratio R of outgoing to return (quick to slow) swings as

$$R = \frac{\cos^{-1}(a/d)}{\pi - \cos^{-1}(a/d)} .$$

If the ratio a/d is varied from zero to unity, the ratio R varies from unity to zero. The above mechanism can also form the basis of a dwell mechanism by allowing the roller B to leave the slot when θ reaches $\cos^{-1}(a/d)$, and designing the output member to have several slots which can receive the roller in turn as shown in Fig. 3.8f. The ratio of quick to slow times now becomes the ratio of output rotation to output dwell times with the limitation that the number of slots n must be an integer. This ratio R is now given by

$$R = (n - 2)/(n + 2).$$

If two diametrically opposite pins are used, R is given by $R = n/2 - 1$.

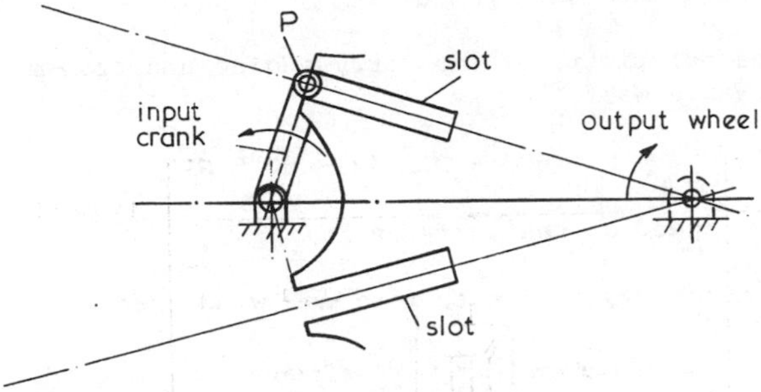

Fig. 3.8f. Dwell mechanism

It must be remembered that the slotted output wheel of Fig. 3.8f rotates in one direction only, with dwell periods interposed. An angular *rocking* motion with an interposed dwell can be achieved by employing a gear wheel, integral with the slotted output wheel meshing with a further gear wheel of the required tooth ratio. This latter gear wheel can then constitute the driving crank in a further extension similar to the arrangement shown in Fig. 3.8d. There are many other ways of obtaining a dwell and one of the simplest is shown in Fig. 3.8g. Here the output lever rocks up and down with

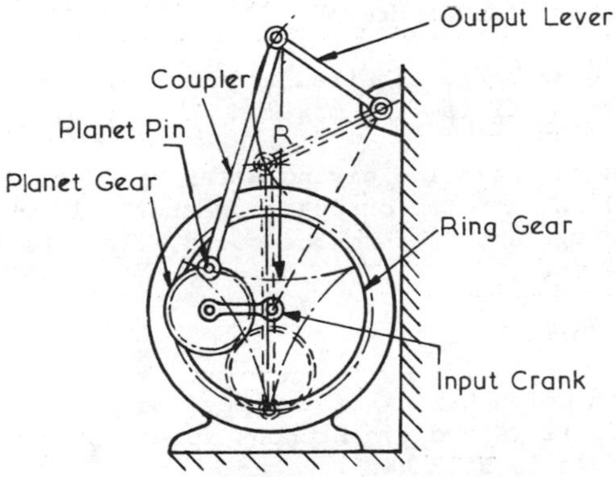

Fig. 3.8g. Dwell mechanism using planet gear

a long dwell in the extreme upper position. The input crank drives a planet gear whose pin traces (in this case) a three-lobed curve, each lobe being an almost circular arc of radius R. If the coupler is made equal to R, the output lever comes virtually to rest during a third (in this case) of the total crank rotation. The output lever then reverses, stops instantaneously in the lower position, reverses and repeats the dwell. By using planet gears of different diameters, different dwell proportions can be obtained.

3.9 Examples

1. A lorry-driver wishes to get more speed out of his lorry so he changes the rear-axle ratio so that, for a given engine-speed, the road-wheel speed is doubled. He finds, however, that the maximum speed is less with the new rear-axle than with the old.

   The output characteristic of the lorry engine and transmission with the original rear-axle was

   $$F = 600 - 0.2\ (V - 50)^2\ N$$

   where V is the speed of the lorry in km/h.

   The demand characteristic due to road and wind drag was

   $$F = 0.06\ V^2\ N.$$

   Explain the inferior performance and find the best axle-ratio and the highest speed possible from the lorry. Assume the same efficiency of power transmission whatever the axle-ratio and that the demand characteristic remains unaltered.

   (Ans. Increase axle speed ratio by factor of 1.23 from the original; 84.5 km/h).

2. A 50 Hz induction motor gives its maximum output of 10 hp (7.46 kW) when it is running at 47 rev/s. It drives a fan whose torque-speed demand characteristic is given by

   $$T_f = 2.8 \times 10^{-3}\ N_f^2\ \text{Nm},$$

   where N_f is the fan speed in rev/s.

   (a) What is the gear-ratio required in the transmission line from motor to fan so as to utilise the maximum motor power?

   (b) A gear ratio is installed giving a fan speed of 1.25 times motor speed. Assuming a motor output characteristic which is linear between 47 rev/s and synchronous speed, find the actual output power of the motor.

   (Ans. (a) 1.6 (b) 5 hp).

3. A 50 Hz induction motor has a torque-speed output characteristic which is linear between 45 rev/s and synchronous speed (50 rev/s). The power delivered at 48 rev/s is 5.13 kW.

   (a) What is the maximum power which can be delivered in the speed range 45 - 50 rev/s?

   (b) The motor drives through a gearbox a machine whose torque-speed demand characteristic is given by $T = 6 \times 10^{-5}\ N^3$ Nm, where N is the machine speed in rev/s. What gear-ratio output speed/input speed is required to deliver 5.13 kW to the load, assuming no losses?

(c) Deduce the equation of the torque-speed demand characteristic seen by the motor for this gear-ratio.

(Ans. (a) 12 kW, (b) 1.265, (c) $T_{RM} = 15.36 \times 10^{-5} N_m^3$.)

4. (a) An analogue of a summing device takes the form shown in Fig. 3.5a view V in which A is driven through a gear ratio from a contra-rotating shaft E (not shown). It is required that three clockwise rotations N_B, N_E and N_C be related in the following way:

$$N_B = - 0.8\ N_E - 1.3\ N_C.$$

Deduce the numbers of teeth required on gears C and E if the numbers of teeth on B and A are 100 and 92 respectively.

(Ans. 130, 32).

(b) What form would a design take to obtain an output speed z given by

$$z = ax + by + cu + dv,$$

where a, b, c, d are constants and x, y, u, v are input speeds?

5. Fig. 3.6b shows a 3-terminal gearbox with terminals at A, B and C. The required speed relationship is given by

$$N_A - 0.14\ N_B + 0.4\ N_C = 0,$$

where N_A, N_B and N_C are in the *specific directions* shown, but the torque directions are not representative of this problem.

(a) Deduce the required tooth ratios C/A and B/X.

(b) When the speeds N_A and N_B are equal to - 500 rev/min and + 1000 rev/min respectively, the resistive torque on A is 300 Nm.

Determine the torques on shafts C and X, stating whether they represent input or output torques with reference to the gearbox.

(Ans. (a) 0.4, 0.1 (b) 120 Nm input, 420 Nm output).

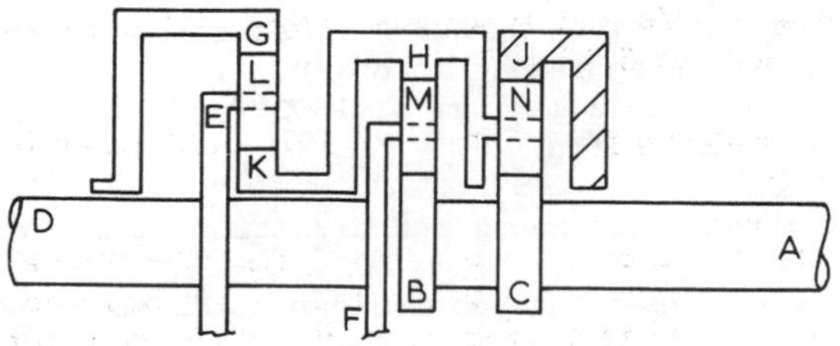

Fig. 3.9a.

6. Fig. 3.9a shows a compound epicyclic gear design. The driving shaft A has two gears, B, C and the driven shaft D has two arms E, F. L, M, N are planet wheels and J is fixed. It is required that the speed ratios N_H/N_A, N_D/N_A and N_G/N_A be 1/5, 1/3 and 11/30 respectively. Calculate suitable values for the ratios

$$\frac{\text{teeth on C}}{\text{teeth on J}}, \quad \frac{\text{teeth on B}}{\text{teeth on H}} \quad \text{and} \quad \frac{\text{teeth on K}}{\text{teeth on G}}$$

If, with these values, a driving torque T_A is applied at A show that the required fixing torque at J is $2T_A$, assuming no losses.

(Ans. 0.25, 0.2, 0.25).

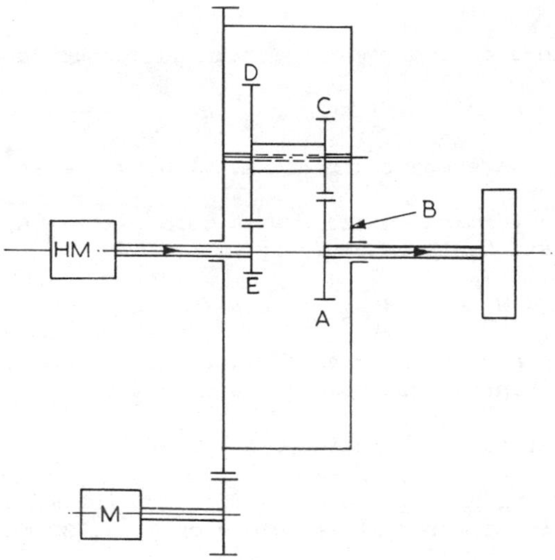

Fig. 3.9b.

7. An infinitely variable stepless speed transmission unit is shown in simplified form in Fig. 3.9b. The numbers of teeth on gears A, C and E are 36, 24 and 24 respectively and all have the same circumferential pitch. The electric motor M drives the member B through a gear at a constant speed of 3000 rev/min under all imposed load conditions. The infinitely variable speed of the driven member A is achieved by driving member E from a small variable speed hydraulic motor HM. The required speed range of member A is from zero to a maximum of 1500 rev/min, and members A and B rotate in the same direction.

What is the required speed range and direction of member E?

(Ans. 375 - 3750 rev/min in opposite direction to that of member A).

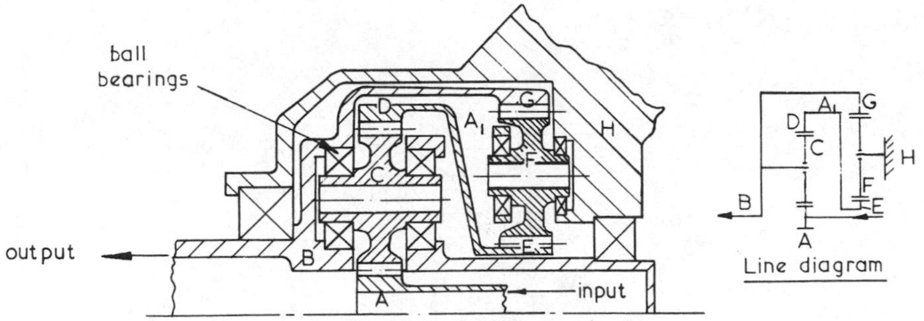

Fig. 3.9c.

8. Fig. 3.9c shows part of a reduction gear used on the Lycoming turbine drive together with its line-diagram. The housing is member H and the input and output power flows are as indicated. Assuming a no-loss system show that the driven to driving torque ratio T_B/T_A is given by

$$T_B/T_A = 1 + D/A\,|1 + G/E|.$$

Find also the fixing torque exerted on the casing as a ratio of the input torque and the ratio of power transmitted by member A_1 to that in the driving shaft.

(Ans. $D/A\ (1 + G/E)$, $D/A\ .\ G/E\ /\ |1 + D/A.(1 + G/E)|$).

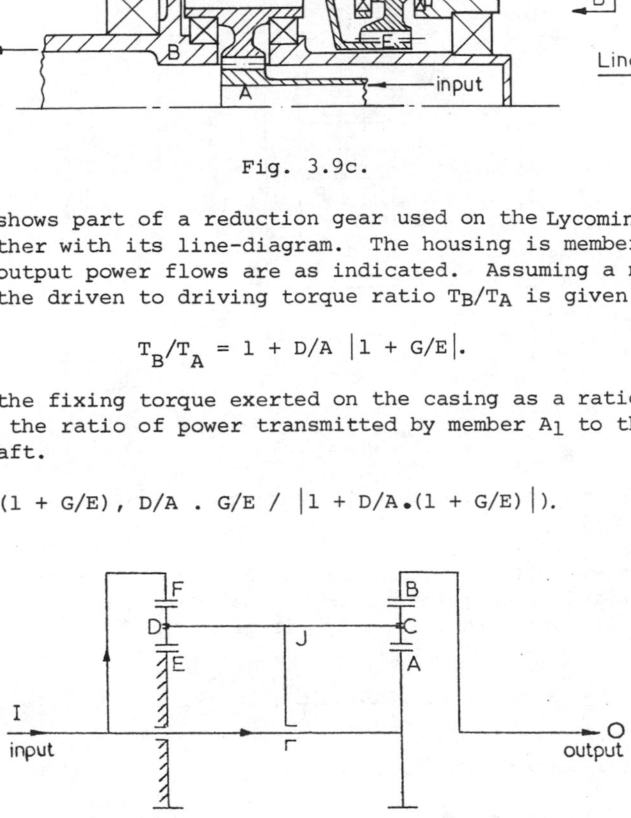

Fig. 3.9d.

9. Fig. 3.9d shows an epicyclic gear train known as the 'Power-flite' transmission in which gear wheel E is held fixed and the input is split between ring gear F and sun gear A. Show that the speed ratio N_O/N_I is given by

$$\frac{N_O}{N_I} = \frac{1 - \frac{EA}{FB}}{1 + \frac{E}{F}},$$

where the capital letters denote numbers of teeth.

Assuming a no-loss system deduce the ratio of power transmitted through gear A to the total power transmitted.

(Ans. $\frac{A}{B}(1 + \frac{E}{F})/(\frac{EA}{FB} - 1)$.)

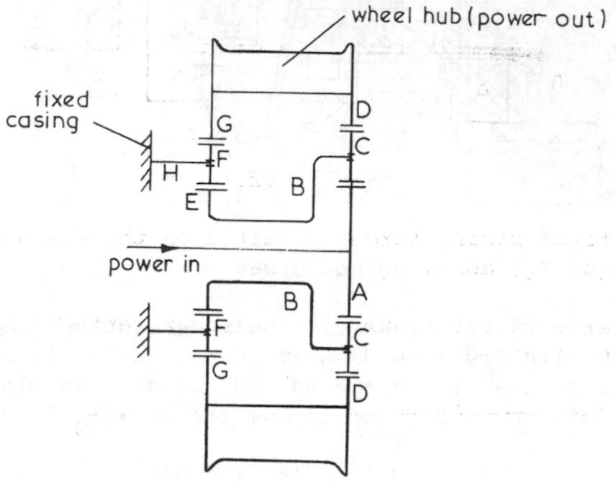

Fig. 3.9e.

10. For what applications do epicyclic gears have significant advantages over ordinary gears?

    Fig. 3.9e shows a line diagram of part of a tractor drive. It comprises an epicyclic reduction gear carried in the wheel hub. Show that, assuming no losses, the torque ratio T_{out}/T_{in} is given by

$$\frac{T_{out}}{T_{in}} = -\left|\frac{D}{A} + \frac{G}{E}\left(1 + \frac{D}{A}\right)\right| ,$$

    where capitals denote numbers of teeth.

    Find also the torque reaction at the fixed casing and the proportion of power transmitted by member B.

    (Ans. $T_A\ (1 + D/A)(1 + G/E)$; $\{|1 + \frac{D}{A}|\ \frac{G}{E}\}/\{|1 + \frac{D}{A}|\ \frac{G}{E} + \frac{D}{A}\}$).

11. Figure 3.9f shows an epicyclic gear in which arm H and wheel A are fixed to the casing.

    (a) Show that the gear ratio N_G/N_C is given by

$$\frac{N_G}{N_C} = \frac{1 + \frac{A}{B}}{\frac{E}{D}\left(1 + \frac{G}{F}\right) + 1} .$$

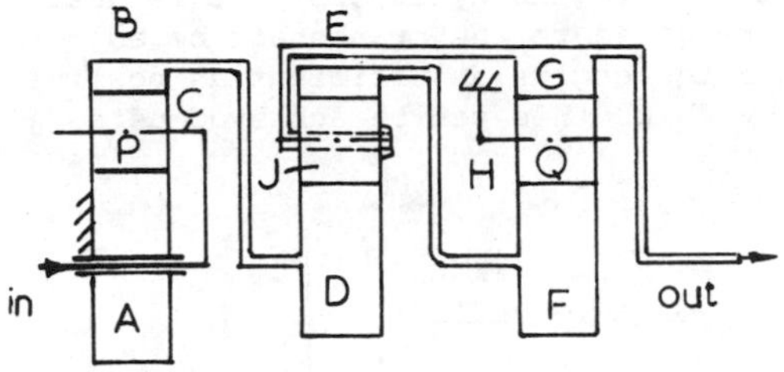

Fig. 3.9f.

(b) Find the total fixing torque required on the casing in terms of the input torque T_i, assuming no losses.

(c) Find in terms of tooth numbers the power ratio $T_E/N_E/T_{out}N_{out}$, assuming no losses.

(Ans. (b) $T_i \left| 1 - \frac{(1 + E/D\ (1 + G/F))}{1 + A/B} \right|$ (c) $\left| 1 + \frac{F}{G} (1 + \frac{D}{E}) \right|^{-1}$).

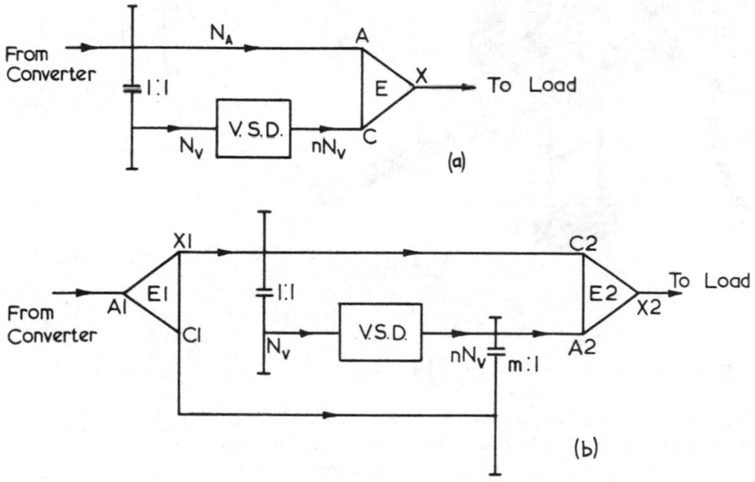

Fig. 3.9g.

12. Fig. 3.9g shows two alternative methods of using a variable speed device (VSD) in a system which is required to pass more power than can the V.S.D. alone. E, E1 and E2 are identical epicyclic gear cells of the form shown in Fig. 3.6b. Show that for a V.S.D. which can pass a given amount of power at a given input speed N_V and velocity ratio n, the power transmitted through system (b) can be greater than that through system (a) for the same overall velocity ratio (as obtained by selection of the ratio m).

13. The crankshaft of an engine drives a camshaft via a chain and sprocket wheels. The camshaft carries cams which open and close the inlet and exhaust valves to the engine cylinders. It is desirable that the phasing of the camshaft relative to the crankshaft be adjustable during running so as to maintain the engine as efficient as possible under all conditions. Devise a means by which this can be achieved using a 3-terminal transmission element.

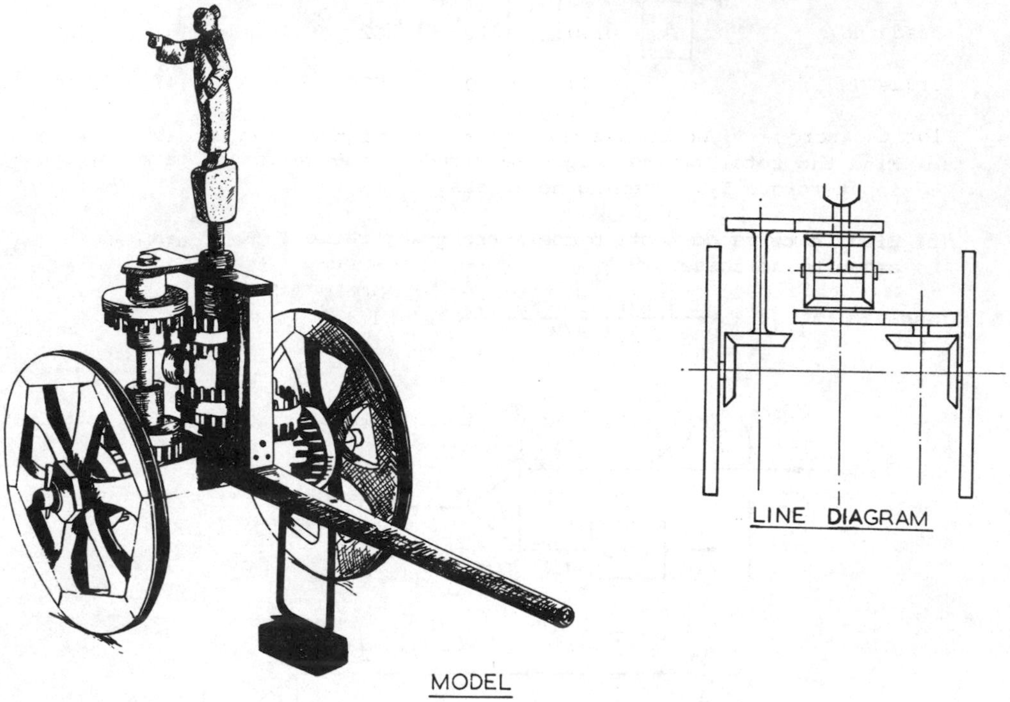

Fig. 3.9h.

14. Fig. 3.9h shows a gear used by the ancient Chinese about 2600 B.C. in the 'South-pointing Chariot'. It was apparently used in crossing the Gobi Desert and can be set so that the figure points south and continues to point south regardless of the direction in which the chariot is going.

    Show that this is indeed possible and deduce suitable numbers of teeth on the gears in terms of the other governing parameters.

15. The characteristics of a torque converter are shown in Fig. 3.3a. For any particular speed ratio N_o/N_i the demand torque of the impeller member varies as the square of the impeller speed N_i and the following table holds:

$\frac{\text{output speed}}{\text{input speed}}$	0	0.2	0.4	0.6	0.8	0.9	0.95
$\frac{\text{input speed rad/s}}{\sqrt{\text{input torque Nm}}}$	1640	1650	1700	1810	2070	2330	2800

The torque-converter is connected to an engine whose torque-speed output characteristic, taking account of torque-converter windage losses, is given by the following pairs of points:

speed rad/s	190	210	262	314	367	420
torque Nm	149	150	152	151	147	140

Plot the torque-speed output characteristic of the torque-converter.

16. A torque-divider takes the form shown in Fig. 3.6c. The ring gear A has 49 teeth and the sun gear C 30. Working from first principles, deduce the torque carried by the coupling as a proportion of the total torque transmitted.

(Ans. 30/79).

17. A pair of contour cams whose centre distance is d is to be used to generate the function

$$\theta_2 = \sin \omega t + k\, \omega t,$$

where ω is the input angular velocity and $\dot{\theta}_2$ the output angular velocity. Deduce the equations for the radii of the input and output cams, working from first principles.

(Ans. $R_1 = \dfrac{d\,(k + \cos \omega t)}{1 + k + \cos \omega t}$, $R_2 = \dfrac{d}{1 + k + \cos \omega t}$).

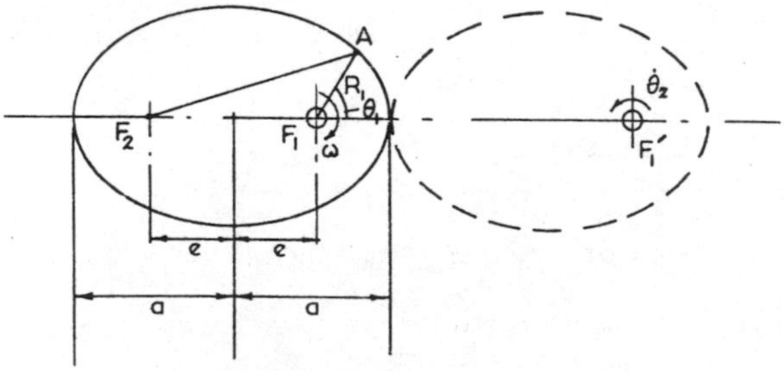

Fig. 3.9i

18. What are the purposes for which non-circular gears are used in a transmission system and what precautions must be taken in their design?

A non-circular gear is in the shape of an ellipse (Fig. 3.9i). It rotates about the point F_1 in a clockwise direction at constant angular velocity ω.

It is a property of an ellipse that $R_1 + F_2A$ = a constant. Show that a second identical ellipse (shown dashed) will rotate in continuous contact with the first.

The polar equation of an ellipse may be written

$$R_1 = \frac{a\,(1 - \varepsilon^2)}{1 + \varepsilon \cos \theta_1},$$

where $\varepsilon = e/a$.

Show that for no slipping the instantaneous output angular velocity $\dot{\theta}_2$ is given by

$$\dot{\theta}_2 = \omega \left| \frac{1 - \varepsilon^2}{1 + \varepsilon^2 + 2\varepsilon \cos \theta_1} \right|.$$

$$P_L = K_i N_i + K_o N_o, \text{ say.}$$

en $$T_{RC} N_i = T_o N_o + K_i N_i + K_o N_o,$$

ving $$T_{RC} = T_o N_o/N_i + K_i + K_o N_o/N_i.$$

so, for true equivalence,

$$\tfrac{1}{2}J_{RC} N_i^2 = \tfrac{1}{2}J_i N_i^2 + \tfrac{1}{2}J_o N_o^2, \qquad (4.2.1)$$

ence $$J_{RC} = J_i + J_o (N_o/N_i)^2.$$

investigating the acceleration of the system, we may now apply Newton's cond law in the form

$$T_C - T_{RC} = J_{RC}\, \delta N_i/\delta t \text{ Nm},$$

ere T_C is the converter output torque. We then proceed as indicated in ection 4.1, remembering that the torque-speed demand characteristic must be onsidered in axes of referred torque T_{RC} and converter speed N_i (not demand orque T_o and demand speed N_o). A similar procedure holds for a multi-terminl gearbox in which the effects of all output shafts are simply added algebracally to give

$$T_{RC} = T_{01} \frac{N_{01}}{N_i} + T_{02} \frac{N_{02}}{N_i} + \ldots\ldots$$

nd $$J_{RC} = J_i + J_{01} \left(\frac{N_{01}}{N_i}\right)^2 + J_{02} \left(\frac{N_{02}}{N_i}\right)^2 + \ldots\ldots,$$

here the suffices 1 and 2 refer to the various outputs.

There is no reason, of course why torques and inertias should not be eferred to the *load*, provided it is realised that the resulting discussion is n terms of load speed rather than converter speed. Fig. 3.1b illustrates a ase in point, where the torque and speed of an automotive engine have been eferred to the load by their conversion to tractive effort and forward speed espectively.

In systems embodying epicyclic gears the translational kinetic energy of ny planet gears must be taken into account in deriving a referred inertia nd equation (4.2.1) would then become

$$\tfrac{1}{2}J_{RC}N_i^2 = \tfrac{1}{2}J_i N_i^2 + \tfrac{1}{2}J_o N_o^2 + \tfrac{1}{2}n\,|m_p v_p^2 + J_p N_p^2|$$

here n is the number of planet gears,
m_p is the mass of each planet gear,
J_p is the moment of inertia of each planet gear,
v_p is the velocity of the planet gear centres,
N_p is the angular velocity of each planet gear,

nd J_i or J_o includes the inertia of the arm carrying the planet gears, epending on whether it is connected to the input or the output of the gearbox.

CHAPTER 4
DYNAMIC PERFORMANCE AND ENERGY STORAGE

Up to now consideration has been given to the performance of a mechanical engineering system when the load or demand torque for example is a function of speed only, and we have discussed the design of a transmission element to ensure adequate matching between a converter and its load under such a condition. In this chapter we shall consider the performance of a system up to its steady state matching point, the requirement for stability of the matching point, the performance of a system under time-fluctuating load conditions and the need for energy storage.

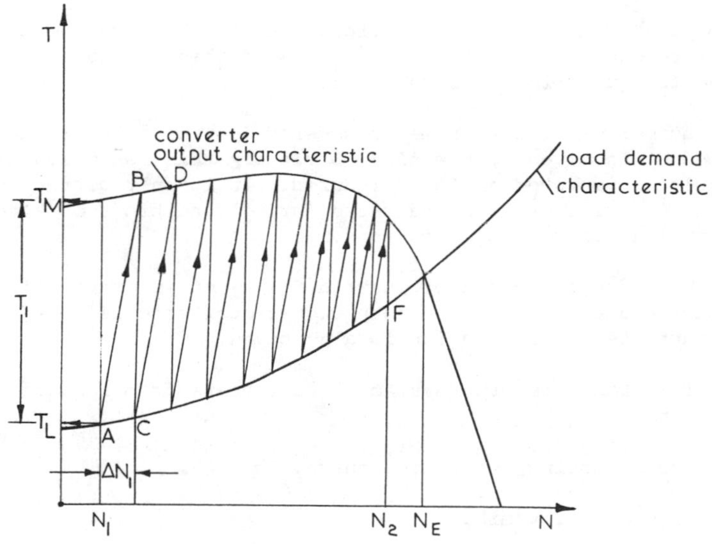

Fig. 4.1a. Acceleration of a converter and its load

4.1 Acceleration of a Converter and its Load

Consider Fig. 4.1a which shows the output characteristic of a converter and the demand characteristic of its load. Assume that the converter is a motor say, which has just been switched on and is in the process of accelerating its load up to equilibrium speed N_E. Further assume that at some instant the speed has reached N_1 at which the motor output torque is T_M and the demand torque as seen by the motor output shaft is T_L. Then, using Newton's second law,

$$T_M - T_L = J\ \Delta N/\Delta t \text{ Nm}, \tag{4.1.1}$$

where J is the moment of inertia of the load in kg m^2 as seen by the motor output shaft and N is the speed of the motor shaft in rad/s.

Thus $$\Delta t = J\ \Delta N/T,$$

where $$T = T_M - T_L.$$

Let it be required to find the time taken to reach speed N_2 from speed N_1. The procedure is then as follows,

1) Select initial speed-increment ΔN_1.

2) Construct line AB (Fig. 4.1a).

3) Calculate $\Delta t = J\Delta N_1/T_1$.

4) Return to point C and repeat with lines such as CD (parallel to AB) until point F is reached. Since the slope, $T/\Delta N$ of these lines is constant they correspond to equal time increments, Δt.

5) Count the number of time increments covered (say n). Then time taken to reach speed N_2 is given by $t = n\Delta t$. Obviously this is an approximate method whose result is improved as Δt is reduced. It is only correct provided the characteristic curves remain unaltered during conditions of acceleration, a fair assumption for most applications.

Worked example A fan of moment of inertia 0.2 kg m^2 is driven directly by an induction motor whose torque-speed output characteristic is given in Fig. 4.1b. The load characteristic for the fan is also given.

(a) Find the time taken for the system to accelerate from rest to a speed of 2650 rev/min.

(b) Plot the corresponding speed-time curve.

Solution (a) with N in rev/min,

$$T_M - T_L = (2\pi/60)\ J\ (\Delta N/\Delta t).$$

For convenience put

$$\frac{\Delta N}{T_M - T_L} = \frac{200}{20} = 10 \text{ (rev/min)/Nm}.$$

Thus $$\Delta t = 10\ (2\pi/60)\ 0.2 = 0.21 \text{ s}.$$

From Fig. 4.1b the number of Δt increments = 12.

Hence total time taken to reach 2650 rev/min = 12 x 0.21 = 2.52 s.

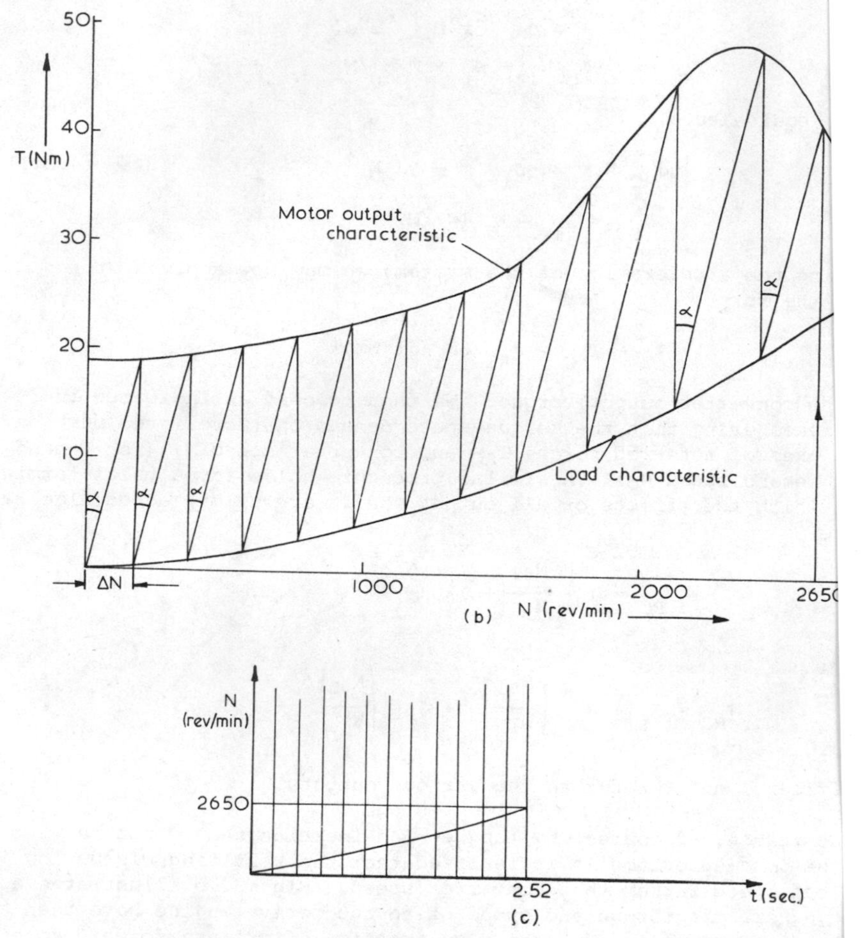

Fig. 4.1b, c. Acceleration and speed of motor

(b) From Fig. 4.1b the ΔN increments corresponding to the Δt increm measured and a plot made of N against t. This is shown in Fig. 4.1

4.2 Referred Torques and Referred Moments of Inertia

In section 4.1 reference was made to a load torque and a load in *seen by* a motor output shaft. Such a torque and inertia may also b be referred to the motor output shaft. Their values depend on any which may be interposed in the transmission between the converter a

Consider a two terminal gearbox whose output torque is T_o. Let of inertia of all the rotating parts connected to the gearbox output (that is the load shaft) and to the gearbox input shaft (the convert be J_o and J_i respectively. Further, let the torque and total moment referred to the converter shaft be T_{RC} and J_{RC} respectively. Let an mission power losses P_L be represented by

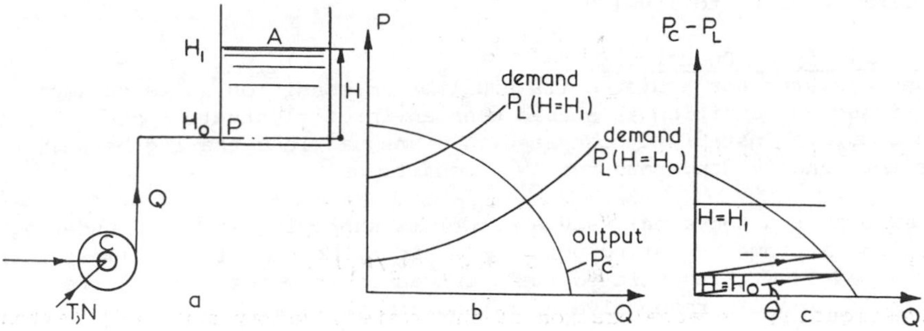

Fig. 4.3. Pump-tank system

4.3 Performance of a Fluid System in reaching its Operating Point

Consider now the case of a centrifugal pump which is in the process of filling a reservoir through a pipe as shown in Fig. 4.3a. Let the output and demand characteristics be as shown in Fig. 4.3b. The lower demand characteristic applies to the case when the level, H in the reservoir is equal to some lower limit H_O and the higher when H equals some higher limit H_1. It is required to evolve a graphical method whereby we can determine the time taken to fill the reservoir from height H_O to height H_1. The problem is similar to that of a motor accelerating its load from a speed N_O say to a speed N_1.

The first observation we can make is that the head H in the reservoir is related to the pressure, P at its base by the equation

$$P = \rho\, g\, H, \tag{4.3.1}$$

where ρ is the density of the liquid.

The pressure P is in turn given by the pump outlet pressure P_C less the pressure loss, P_L incurred by the fluid in reaching the reservoir.

Thus

$$P = P_C - P_L. \tag{4.3.2}$$

Using equations (4.3.2) and (4.3.1) and differentiating with respect to time gives

$$\frac{\Delta\,(P_C - P_L)}{\Delta t} = \rho\, g\dot{H} = \frac{\rho g}{A}\,.\,Q,$$

where A is the surface area of the reservoir, assumed constant.

Proceeding in a fashion similar to the procedure of section 4.1 we may also write

$$\frac{\rho g}{A}\,.\,\Delta t = \frac{\Delta\,(P_C - P_L)}{Q}\,.$$

It now remains to plot a graph of $P_C - P_L$ against Q (Fig. 4.3c) in which each element is constructed by selecting a time interval, Δt and calculating

$$\frac{\rho g}{A}\,.\,\Delta t = \frac{\Delta\,(P_C - P_L)}{Q} = \tan\,\Theta.$$

Adding the time intervals then gives the total time needed to fill the reservoir from level H_o to level H_1.

4.4 Stability of Operation

When a system has achieved its equilibrium condition it is of importance to know if such a condition is stable when an input fluctuation or a load fluctuation occurs. Instability can result in unequal load-sharing between parallel converters and in dangerous run-away conditions.

Assume that a converter (say a motor) is supplying its load under equilibrium conditions at some operating point. Consider the effect of a small increase in motor speed ΔN with associated small increases in motor and referred load torques ΔT_m and ΔT_L respectively. Then we may write,

$$(T_m + \Delta T_m) - (T_L + \Delta T_L) = J \frac{\delta}{\delta t} (N + \Delta N),$$

where J is the total moment of inertia referred to the motor.

But
$$T_m - T_L = J\, \delta N/\delta t$$

and thus
$$\Delta T_m - \Delta T_L = J \frac{\delta}{\delta t} (\Delta N).$$

If T_m and T_L are functions of motor speed only, then

$$(\partial T_m/\partial N)\ \Delta N - (\partial T_L/\partial N)\ \Delta N = J \frac{\delta}{\delta t} (\Delta N)$$

or
$$T'\Delta N = J \frac{\delta}{\delta t} (\Delta N),$$

where
$$T' = (\partial T_m/\partial N) - (\partial T_L/\partial N).$$

Thus
$$\int dt = \frac{J}{T'} \cdot \int \frac{d(\Delta N)}{\Delta N},$$

giving
$$\Delta N = k_{18} \exp (T't/J),$$

where k_{18} is a constant of integration.

Thus, provided T' is negative, ΔN will vanish in time and the system will return to equilibrium. Hence for stability we require that $\partial T_m/\partial N < \partial T_L/\partial N$ at the matching point. Equilibrium points such as S_1 (Fig. 4.4a) are thus more stable than points such as S_2.

Instability can be explained physically by considering points such as U in Fig. 4.4a. Suppose that for some reason the discharge Q of a pump say increases slightly; then the net pressure causing flow will also increase, thus increasing the flow still further and the result is an instability in the delivery line. It is also possible for a pump to give two different discharge rates Q for a given value of pressure P. This can lead to pulsations or surging which are accentuated if two such pumps are run in parallel. If the pump is run at a point S however, any increase in flow results in a decrease in the net pressure causing flow and hence an automatic flow reduction and this constitutes a stable state of affairs.

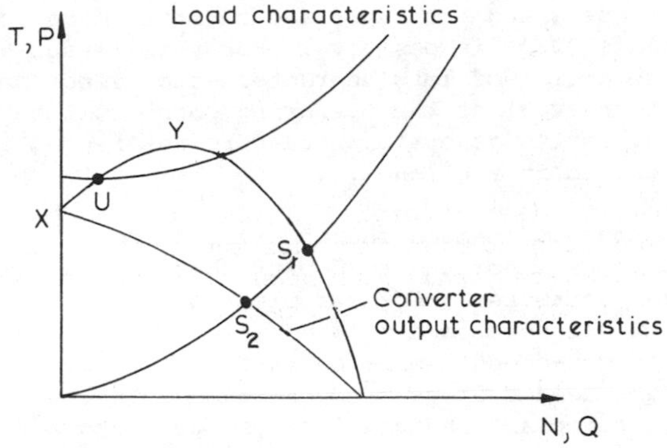

Fig. 4.4a. Stability of operating point

For rotodynamic air compressors the characteristics embody an even more pronounced rising portion XY and the surging effects are more noticeable. There is an additional limitation to the operating range as choking occurs in the diffuser throat when Q reaches a certain value. Choking is a limiting of the flow owing to a limiting velocity with which a gas may be expelled. This limiting velocity corresponds to the speed of sound, being the speed with which pressure alterations make themselves felt upstream. Any reduction of pressure below some critical value (corresponding to a gas velocity equal to the speed of sound) cannot be conveyed as information upstream against the flow and the supply does not therefore respond by supplying a greater flow. Points

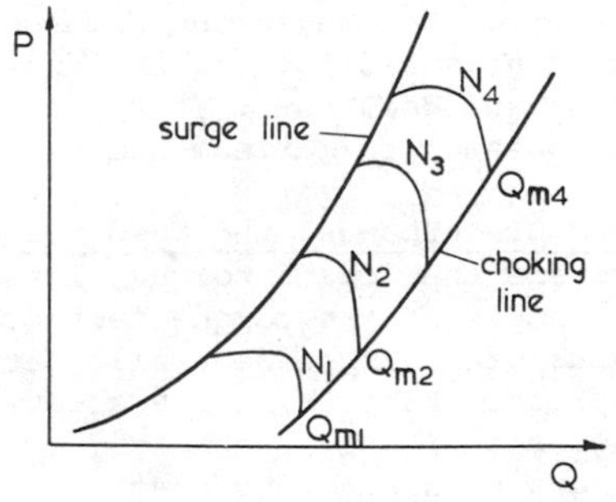

Fig. 4.4b. Rotodynamic compressor characteristics

such as Q_M (Fig. 4.4b) represent the maximum deliveries obtainable at the particular rotational speeds N for which the curves are drawn. The actual variation of pressure over the complete range of mass flow and speed will be shown by curves such as those in Fig. 4.4b. The left-hand extremities of the constant speed curves may be joined to form what is known as the 'surge line', while the right-hand extremities represent the points where choking occurs.

Consider now an electrical machine which could exhibit instability. The series generator discussed in section 2.2 has for its voltage-current output equation

$$V = I\,(k_7\,N - R).$$

Thus, for a typical input speed N the voltage-current characteristic has a positive slope and hence $\partial V/\partial I$ is positive. For this reason great care must be exercised in the selection of load characteristic, since one having a slope whose value is less than $\partial V/\partial I$ at the operating point will cause instability. This is the reason why series generators are very restricted in their use to special loads as discussed in section 2.1.

When inspecting output and demand characteristics for the purpose of assessing stability, it is important to depict these characteristics in terms of the proper constant parameter. For example, in the above case of the series generator, it might be argued from equation (2.2.9) that for constant output voltage, the torque-speed demand characteristic is conducive to instability since the torque demand drops with increase in speed. However, output voltage is a rather artificial constant for most generators. The true demand characteristic which a generator-drive sees is obtained only with a knowledge of the load imposed on the generator. For example, let us assume that the generator feeds power into a constant resistance. The voltage-current demand characteristic seen by the generator is thus a straight line of positive slope from which a succession of pairs of values of voltage and current can be read. Referring now to equations (2.2.8) and (2.2.9), we may deduce the torque and speed corresponding to the first pair of values, and these will give a point on the true (constant-resistance) demand characteristic seen by the generator-drive. Similar points may be obtained for the other pairs, and so the true demand characteristic may be constructed. Such a demand characteristic was obtained in the first worked example of section 2.2.

A potential stability hazard exists for all self-sustaining converters, since these have no starting-torque, voltage or pressure and the slope of the output characteristic must necessarily initially rise to some maximum value, that is, have a positive slope. In many cases, however, the load characteristic seen by the converter starts at the origin of the axes of the potential parameter and the rate parameter and does not meet the converter output characteristic until the latter has passed its maximum and has a negative slope.

4.5 Specific Forms of Load Disturbance: the Need for an Energy Store

In many engineering systems the demand torque, instead of being steady, is periodic in nature. One example is the torque demanded by a punching machine because of the sudden spurts of power needed while its punch is shearing through the material of the work-piece. Another is the torque demanded at the input crank of a load consisting of a linkage mechanism, even though this mechanism may be ultimately supplying a constant torque at its output. The converter of such a system is often of a type which operates best under conditions of constant torque and speed. Hence we are faced with a matching problem of a new type, where the supply torque and the referred demand torque are different at different points in time and where the latter may at some times be greater than the former. The question arises as to what element could be used now in the transmission to enable the converter and the load to be matched. The answer is an energy *store* which has the ability to receive and supply energy as needed.

Imagine a converter and its load connected by an element which transmits torque, fluid pressure and heat (Fig. 4.5a). The first law of thermodynamics tells us that, for the control volume shown, the heat plus work supplied to the control volume is equal to the heat plus work taken out of the control volume plus the *increase* in internal energy in the control volume. Up to now

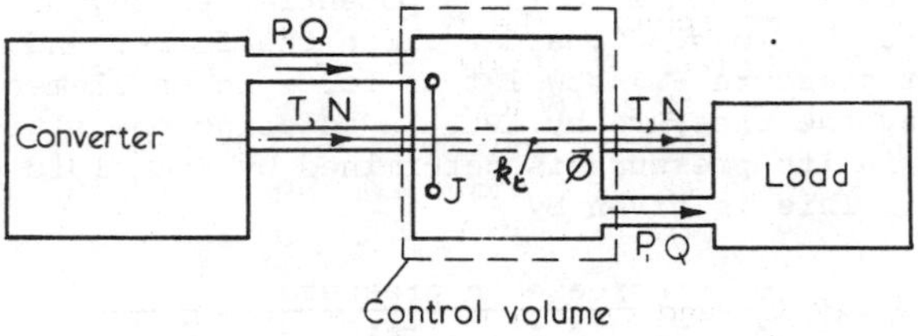

Fig. 4.5a. General storage element

we have assumed that the last term has only comprised kinetic energy of rotation but of course this is a restriction and in general all forms of energy should be included. Such energy constitutes energy of *storage* and is of prime importance in the present context. Consider first the mechanical link transmitting torque and speed. This link contains the kinetic energy of rotation already referred to by virtue of the presence of some rotating body of inertia J, which is inherent in any mechanical link, but which may be augmented by the provision of a fly-wheel. If the converter torque just to the left of J (that is at the left hand edge of the control volume) is T_c and the demand torque at the right hand edge of the control volume is T_L, then we may write Newton's second law as

$$T_c - T_L = T = J\, dN/dt = JN\, .\, dN/d\theta,$$

where θ is the angle turned through by the rotating body of inertia J and N is equal to $\dot{\theta}$. Hence the kinetic energy ε_k stored by the rotating body is equal to the work done by the net torque T on the inertia, J. That is,

$$\varepsilon_k = \int_0^\theta T d\theta = \int_0^N JN\, .\, \frac{dN}{d\theta}\, .\, d\theta = \tfrac{1}{2}JN^2. \qquad (4.5.1)$$

The mechanical link will also twist as a result of the demand torque imposed on it and in so doing will acquire some strain energy. If the twist is assumed to be concentrated in the load shaft of torsional stiffness k_t then the strain energy ε_t stored in this shaft is equal to the work done on the shaft by the torque T_L which is equal to the driving torque T_c under steady-state conditions.

Thus
$$\varepsilon_t = \int_0^\phi T_L\, d\phi,$$

where ϕ is the angle of twist produced in the load shaft.

Now
$$T_L = k_t\phi$$

and so
$$\varepsilon_t = k_t \int_0^\phi \phi\, d\phi = \tfrac{1}{2}k_t\phi^2. \qquad (4.5.2)$$

Strain energy can be augmented in an engineering system by the provision of a specially designed torsion spring, a good example being the use of a spiral spring in a clock mechanism.

Consider now the fluid link in Fig. 4.5a which, by virtue of fluid compressibility and the height difference between inlet to the control volume and outlet from the control volume contains potential energy due to pressure and to height relative to the load. Consider first the former and let the control volume be ν under pressure P. Now let us force in an elemental volume $\Delta\nu$, thereby increasing the pressure by ΔP. In so doing the fluid will be compressed and the increase in its pressure is determined by the fluid property known as bulk modulus, K_B. This is given by

$$K_B = \frac{\text{increase in pressure}}{\text{volumetric strain}} = \frac{\Delta P}{\Delta\nu/\nu}$$

or

$$K_B = \nu \cdot \frac{\Delta P}{\Delta\nu}. \tag{4.5.3}$$

But the inflow rate Q is given by

$$Q = \frac{\Delta\nu}{\Delta t},$$

and so

$$K_B = \frac{\nu}{Q} \cdot \frac{\Delta P}{\Delta t}. \tag{4.5.4}$$

The pressure energy ε_p stored in the control volume is equal to the work done by the inflow acting against the pressure P inside the control volume. That is

$$\varepsilon_p = \int PQ\,\Delta t = \int P\,\frac{\nu}{K_B}\,\Delta P.$$

For a liquid the bulk modulus K_B may be assumed to be independent of pressure and the effect of any increase in temperature may be ignored. Hence we may say that

$$\varepsilon_p = \frac{\nu}{K_B}\int_0^P P\,\Delta P = \tfrac{1}{2} \cdot \frac{\nu}{K_B} \cdot P^2 \tag{4.5.5}$$

For a gas the above assumption may not be made. However, we may note that, whether the fluid be a liquid or a gas, equation (4.5.3) may be used. Assuming that the volume $\Delta\nu$ forced in increases the density ρ such that

$$\frac{\Delta\nu}{\nu} = \frac{\Delta\rho}{\rho},$$

then we may further write

$$K_B = \rho\,\frac{\Delta P}{\Delta\rho} \tag{4.5.6}$$

Now, for a gas, including air, we may assume that

$$\frac{P}{\rho^n} = \frac{P + \Delta P}{(\rho + \Delta\rho)^n},$$

where n is a constant index which depends upon the rapidity with which the gas is compressed. The two extreme values for n are 1 (for isothermal compression) and about 1.4 for adiabatic compression, that is compression with no associated heat gain or loss by the contents of the pressure vessel. For $\Delta\rho/\rho << 1$, as occurs with most pressure vessel applications as energy stores, then we may approximate by writing

$$\frac{P}{\rho^n} \simeq \frac{P + \Delta P}{\rho^n \left|1 + \frac{n\Delta\rho}{\rho}\right|}$$

whence

$$\rho\frac{\Delta P}{\Delta\rho} \simeq Pn.$$

Hence, from equation (4.5.6),

$$K_B \simeq Pn \qquad (4.5.7)$$

and

$$\varepsilon_p = \int_o^P \frac{P\nu}{Pn} \,.\, \Delta P = \nu P/n.$$

If the control volume were in the form of a storage tank or reservoir in which it were required to store energy by virtue of liquid head H, then we may write,

$$\text{base pressure } P = \rho g H$$

and

$$\text{inflow } Q = A\dot{H},$$

where A is the surface area of the liquid in the tank. Hence the energy ε_h stored in the reservoir is equal to the work done in introducing the liquid. That is

$$\varepsilon_h = \int_o^t PQdt$$

$$= \int_o^P \frac{AP}{\rho g} \,.\, dP$$

or

$$\varepsilon_h = \tfrac{1}{2} \,.\, \frac{A}{\rho g} \,.\, P^2, \qquad (4.5.8)$$

assuming the area A remains constant.

The electrical counterpart of the flywheel is the inductance coil. Such a coil has the property of inductance which opposes any change in current in an electrical system, just as inertia opposes any change in speed in a mechanical system. The inductance induces a back-e.m.f. to oppose any change in current, whatever the direction of the change. While the current is increasing, energy is being stored in the associated magnetic field as it is built up around the inductance. When the current decreases this energy is given up by the magnetic field. The greater the number of adjacent turns of wire cut by the magnetic field the greater the back-e.m.f. induced, and a coil having many turns has a high inductance. An iron core inserted in the coil increases the inductance by concentrating and increasing the magnetic field near the coil. The basic unit for inductance is the henry,and an inductance coil is of value one henry if the back-e.m.f. in it is one volt when the current changes at the rate of one ampere per second.

Thus $$V = L\dot{I} \text{ volts},$$

where L is the inductance in henrys and I is the current in amperes. Hence L may be seen as linking the rate parameter, with the time integral of the potential parameter, just as did the flywheel. The energy ε_L stored in the inductance is given by

$$\varepsilon_L = \int V dq,$$

where q is the charge, being the integral of current I with respect to time t. Thus $\dot{q} = I$ and the energy stored,

$$\varepsilon_L = \int_0^t V\ I\ dt = \int_0^I LIdI = \tfrac{1}{2}LI^2.$$

There are also many electrical circuits which tend to resist changes in voltage and are said to contain capacitance. A capacitor consists basically of a pair of charged plates and its unit of measure is the farad. Factors which affect the capacitance are the insulating material or dielectric between the charged plates, the plate area and the distance of separation between the plates. The potential created between the plates of a capacitor depends upon the charge introduced which is the time integral of the current I.

Thus $$V = q_C/C = \frac{1}{C}\int Idt.$$

The energy ε_C stored in the capacitor is equal to the work done in introducing the charge q_C against the potential V.

Thus $$\varepsilon_C = \int_0^V q_C dV = \int_0^V CVdV$$

or $$\varepsilon_C = \tfrac{1}{2}CV^2. \tag{4.5.9}$$

In storing potential energy the capacitor is analogous to the torsion spring, the pressure vessel and the liquid reservoir.

All the expressions for energy stored are deduced from the basic fact that work has been done to produce this energy, whether by creating acceleration to acquire a final speed, forcing gas into a pressure vessel to produce pressure energy and so on. There is an alternative method of acquiring stored energy, particularly relevant to a gas, and this is by the application of work and heat to produce internal energy by virtue of a temperature rise. Here we shall consider only those thermal energy storage devices which involve no work transfer. Consider then the thermal equivalent of the torsion-shaft. This is the heat-storage capacity of a mass m of material realised by its specific heat C_H. Thus heat flow rate

$$q = mC_H\dot{\psi} \tag{4.5.10}$$

and energy stored

$$\varepsilon_{th} = \int_0^t qdt = \int_0^\psi mC_H d\psi$$

or $$\varepsilon_{th} = mC_H\psi, \qquad (4.5.11)$$

where ψ is the temperature rise.

This energy stored bears a linear relationship to the potential parameter ψ rather than the parabolic relationship met in the other energy expressions. This is because q itself (rather than the product $q\psi$) is a measure of power.

It should be noted that, in recovering energy from a storage device, there is a limit set by the Second Law of Thermodynamics on the proportion of that energy which can appear as work rather than as heat. This proportion depends upon the temperatures of the store and of the demand and determines the theoretical efficiency of the load device in using this energy.

4.6 Limitations to the Different Forms of Energy Storage

An inspection of the various expressions for energy stored in the different devices will indicate that in all cases this energy increases with increase in the rate or in the potential parameter. The question thus arises as to what factors might limit such a parameter and, as a result, what devices are to be preferred on the logical basis of energy stored per unit mass of material.

Proceeding through the mechanical devices in the order discussed we can first consider the flywheel whose energy depends upon its moment of inertia J and its rotational speed N. If the flywheel is in the form of a rim held to the shaft by spokes rather than in the form of a disc, say its moment of inertia will then be a maximum for a given mass m of flywheel material and we may say that

$$J \simeq mr^2,$$

where r is the mean radius of the flywheel rim. Hence the energy stored may be written,

$$\varepsilon_k = \tfrac{1}{2}JN^2 = \tfrac{1}{2}m\,(rN)^2.$$

But clearly the greater the value of centrifugal force the greater the bursting effect on the rim. We may deduce this bursting effect by considering a small element of the rim as shown in Fig. 4.6a. The element will burst away as soon as the tension force T exceeds a certain limit set by the material of the rim.

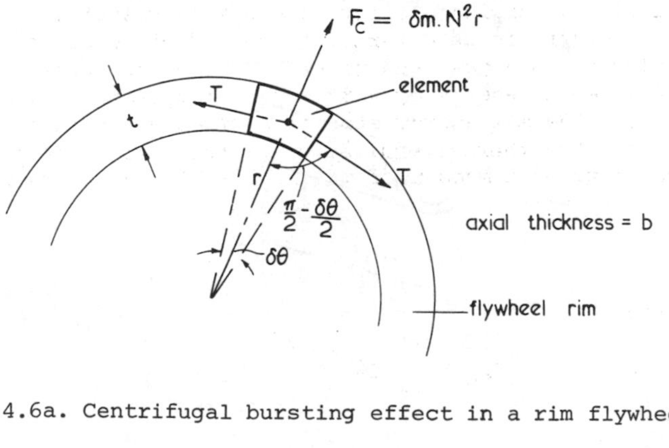

Fig. 4.6a. Centrifugal bursting effect in a rim flywheel

The centrifugal force F_c on the element is given by

$$F_c = \delta m \cdot N^2 r,$$

where N is the rotational speed in rad/s

and $$\delta m = \text{the mass of the element} = \frac{m\delta\theta}{2\pi}.$$

For equilibrium we require that

$$2T \cos(\pi/2 - \delta\theta/2) = m\delta\theta N^2 r/2\pi .$$

Since $\delta\theta$ is small this equation may be rewritten

$$T\delta\theta = m\delta\theta N^2 r/2\pi.$$

That is $$T = mN^2 r/2\pi.$$

Now the tensile stress τ_t due to the tension force T may be written

$$\tau_t = T/bt \simeq 2\pi Tr\rho/m,$$

where ρ is the density of the flywheel material and hence m/ρ is the volume.

Hence $$\tau_t = \frac{2\pi r\rho}{m} \cdot \frac{mN^2 r}{2\pi} = \rho \cdot (rN)^2.$$

Thus, the energy stored, ε_k, may now be written

$$\varepsilon_k = \tfrac{1}{2}m \cdot \frac{\tau_t}{\rho}$$

or $$\varepsilon_k/m = \tau_t/2\rho. \qquad (4.6.1)$$

The permissible tensile stress $\tau_{t_{max}}$ is thus a measure of the maximum mass effectiveness of energy storage. The need for spokes to connect the rim to a central shaft reduces the value given in equation (4.6.1), but the latter serves as an ideal to be aimed at.

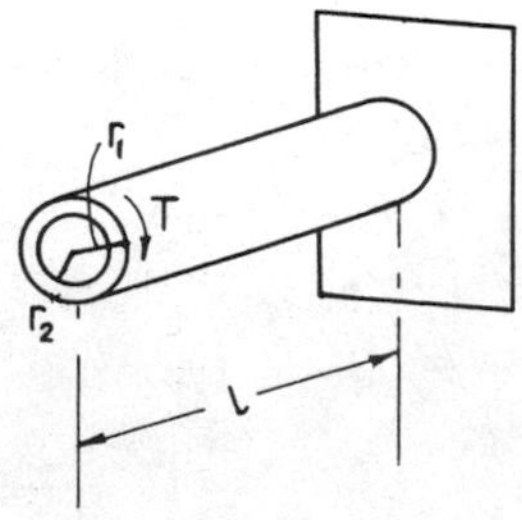

Fig. 4.6b. Torsion spring

Now consider the torsion spring of Fig. 4.6b whose torsional stiffness is k_t Nm/rad. Let the spring be in the form of a tube whose inner and outer radii are r_1 and r_2 respectively. Then we may write

$$T/I = G\phi/\ell,$$

where T is the applied torque, I is the polar second moment of area of the cross section = $(\pi/2)\ (r_1^4 - r_2^4)$, G is the modulus of rigidity of the material and ϕ is the angle of twist over length ℓ.

Thus the torsional stiffness k_t is given by

$$k_t = T/\phi = GI/\ell = \frac{\pi G}{2\ell}\ (r_1^4 - r_2^4).$$

Let us first consider the conditions for achieving a maximum value of the ratio torsional stiffness per unit mass. If the density of the material is again ρ then the mass m of the torsion spring is given by

$$m = \rho\pi\ell\ (r_1^2 - r_2^2)$$

and hence
$$\frac{k_t}{m} = \frac{\pi G\ (r_1^4 - r_2^4)}{2\ell\rho\pi\ell\ (r_1^2 - r_2^2)} = \frac{G}{2\ell^2\rho}\ (r_1^2 + r_2^2).$$

Thus, for a given value of r_1, r_2 should be as large as possible, that is

$$r_2 \simeq r_1.$$

Hence
$$k_t/m \simeq Gr_1^2/\ell^2\rho. \tag{4.6.2}$$

Thus the energy stored per unit mass ε_t/m is given from equation (4.5.2) by

$$\varepsilon_t/m = k_t\phi^2/2m \simeq (G/2\rho)\ (r_1\phi/\ell)^2.$$

Now it may be shown from material strength considerations that the shear stress τ_a induced in the material is given by

$$\tau_a = G\phi r_1/\ell,$$

and so the energy stored per unit mass is

$$\varepsilon_t/m = (G/2\rho)\ (\tau_a/G)^2 = \tau_a^2/2G\rho, \tag{4.6.3}$$

in which τ_a is the shear stress in the material. This has a maximum value of about 200,000 kN/m^2 for steel. The modulus of rigidity G of steel has a value of about 100 x 10^6 kN/m^2.

In many instances a low torsional stiffness k_t is required and it may be seen from equation (4.6.2) that this necessitates a sufficiently large value of axial length ℓ. This is not always so easy to accommodate so an alternative method of strain energy storage is employed in either a helical or a spiral spring, a clock-spring being a good example. These devices store tensile strain energy instead of shear strain energy and corresponding values of energy stored per unit mass are

$$\varepsilon/m = f^2/8\ E\rho$$

and
$$\varepsilon/m = f^2/24E\rho \tag{4.6.4}$$

for the helical and spiral spring respectively. In these equations f is the tensile stress, having a typically maximum permissible value for steel of about 250,000 kN/m^2, and E is the modulus of elasticity of the steel (about 210 x 10^6 kN/m^2). Comparison of equations (4.6.3) and (4.6.4) with equation (4.6.1) shows that a flywheel is a far more effective energy store than is a torsion, helical or spiral spring (when considered on the basis of storage per unit mass) owing to the high values of the moduli E and G.

Now let us consider the fluid energy storage devices in terms of mass effectiveness. For storage of the potential parameter namely pressure we need either a reservoir or a pressurised chamber and no natural alternative exists for storing the relevant rate parameter. The mass m of a cylindrical reservoir is given by

$$m = \rho AH,$$

and so, from equation (4.5.8),

$$\varepsilon_h/m = AP^2/2\rho^2 gAH = P/2\rho,$$

since
$$P = \rho gH.$$

Alternatively we may write

$$\varepsilon_h/m = \rho gH/2\rho = gH/2.$$

For the pressurised chamber containing a liquid it has been shown in equation (4.5.5) that

$$\varepsilon_p/\nu = P^2/2K_B$$

and hence
$$\varepsilon_p/m = P^2/2K_B\rho.$$

For a gas $K_B \simeq nP$ (equation (4.5.7)) and hence $\varepsilon_p/m \simeq P/2n\rho$. Thus the lower the fluid density, the more beneficial will it be to store energy in a pressure chamber rather than in a reservoir. In most cases reduction of fluid height is also a dominating influence and so a pressurised chamber is the more common method of fluid energy storage. However, a limitation is then imposed by the thickness and hence mass of material required by the container to prevent bursting. It then becomes pertinent to consider the energy stored per unit mass of containing material. If the diameter d of the chamber is assumed large compared with the thickness t of its material, then, to a good approximation, we may say that the hoop stress produced in the material as a result of the tendency to burst one semi-circular half from the other along its length l is given by

$$\tau = Pd/2t. \tag{4.6.5}$$

If we further assume that the pressure vessel has a length l considerably greater than its diameter d, as in most practical instances, then the mass m_c of the container may be written

$$m_c \simeq \pi dtl\rho_c,$$

where ρ_c is the density of the vessel material.

Thus the energy ε_p stored per unit mass m_c of pressure vessel is given from equations (4.5.5) and (4.5.7) by

$$\varepsilon_p/m_c \simeq \frac{P\nu}{2nm_c} \simeq P\nu/2n\pi dtl\rho_c.$$

Using equation (4.6.5) this becomes

$$\varepsilon_p/m_c = \tau\nu/\pi d^2 nl\rho_c \qquad (4.6.6)$$

Finally, putting

$$\nu = \pi d^2 l/4$$

gives

$$\varepsilon_p/m_c \simeq \tau/4n\rho_c. \qquad (4.6.7)$$

Hence a pressure vessel material permitting as high a value as possible of hoop stress τ should be used. Since hoop stress τ is a tensile stress and n is about unity for most conditions, comparison of equations (4.6.1) and (4.6.7) shows that a flywheel is about twice as effective as a gas-pressurised chamber in terms of energy stored per unit mass containing material when both operate with maximum tensile stress. This fact can be of considerable importance when designing a complete system in which some choice can be exercised as to the intermediate energy forms.

For comparison purposes consider now means of achieving storage of electrical energy. For a capacitor (the electrical equivalent of the torsion spring), the energy stored ε_c has been shown in equation (4.5.9) to be given by

$$\varepsilon_c = \tfrac{1}{2}CV^2,$$

where C is the capacitance and depends upon the dielectric material of thickness t, lying between two charged capacitor plates of area A. In fact

$$C = k_d A/t,$$

where k_d is the absolute permittivity of the material. Now the maximum voltage which can be sustained across the plates is prescribed by the so-called breakdown voltage V_m which has the units of volts/metre. Thus, when the capacitor is working at its limit,

$$V = V_m t,$$

giving

$$\varepsilon_c = \tfrac{1}{2} \,.\, (k_d A/t) \,.\, (V_m t)^2,$$

whence the maximum possible energy stored per unit mass of dielectric material is

$$\varepsilon_c/m = k_d V_m{}^2/2\rho_d,$$

where ρ_d is the density of the dielectric material.

For a simple inductance coil (the electrical equivalent of the flywheel), the energy ε_L stored has been shown to be given by

$$\varepsilon_L = \tfrac{1}{2}LI^2,$$

where L is the inductance. This depends upon the density B of magnetic flux in its iron core per ampere of current I, flowing in the coil and upon the permeability μ of the core material, such that

$$L = (v/\mu) \cdot (B/I)^2,$$

where v is the volume of iron-core material. B is limited by saturation of the iron, which in turn limits the energy stored, and this becomes

$$\varepsilon_L = (v/2\mu)(B/I)^2 I^2 = vB^2/2\mu,$$

where B is the flux density. Hence the energy stored per unit mass m_c of core material is

$$\varepsilon_L/m_c = B^2/2\mu\rho_c,$$

where ρ_c is the density of core material.

Finally, consider thermal energy storage, for which we may write,

$$\varepsilon_{th}/m = C_H\Psi,$$

indicating that the specific heat C_H and the maximum permissible temperature Ψ to avoid damage to the material or its surroundings are the important factors in determining the maximum energy storage per unit mass.

The following table indicates values of the energy stored per unit mass for various energy stores, when sample values of the controlling parameters are used.

The values in the table should be compared with the mass effectiveness of fossilised fuel and of the electrical lead acid battery, which have rough values of 40,000 and 70 kJ/kg respectively. The relatively high effectiveness of the flywheel as a *mechanical* energy store should also be noted. This has resulted in an ever-increasing utilisation of the flywheel. Until recently a restriction was placed on the use of flywheels by the range of suitable materials available; for example, the table assumes an average steel to be the construction material. This range has greatly increased of late largely due to aerospace technology which has produced fibre composites of high tensile strength and yet of low density. For example, PRD-49, a commercial fibre produced on a fairly large scale, can store seven times as much energy per unit mass as high-strength steel. Such flywheels offer a means for electric utilities to cope with peak loads. In the past peaking units, usually gas-turbine driven generators, were used only during the time when demand was heaviest and capital and fuel costs tended to be high. However, the peaking problem is not caused by a basic energy shortage, provided energy can be stored during periods of low demand. One solution has been to use pumped storage in which water is pumped uphill to a reservoir during off-peak hours and allowed to flow downhill during peak hours, thus powering a hydroelectric generating plant. An excellent example of this is the Ffestiniog Power Station in North Wales which operates with a head difference of 300 metres and an overall efficiency of 73%. Unfortunately, there are few places where pumped storage is geographically practical and the flywheel then offers considerable advantages. Such a flywheel would be coupled to a generator-motor, which would function as a generator when the system was drawing on the energy stored in the flywheel and as a motor when energy was being stored in the flywheel by accelerating it. The

Mass-effectiveness of various energy stores

Device	Energy stored per unit mass	Assumed values of parameters	Approximate max. value of energy stored per unit mass kJ/kg
thermal store	$C_H\Psi$	specific heat C_H = 0.88 kJ/kgK temperature rise Ψ = 500°C	440
flywheel	$\tau/2\rho$	tensile stress $\tau = 2.5 \times 10^8$ N/m^2 material density ρ = 7700 kg/m^3	16
pressure vessel (gas)	$\tau/4n\rho_c$	tensile stress $\tau = 2.5 \times 10^8$ N/m^2 material density ρ_c = 7700 kg/m^3 n = 1	8
liquid reservoir	gH/2	head of liquid H = 200 m	1
torsion spring	$\tau_a^2/2G\rho$	shear stress $\tau_a = 2 \times 10^8$ N/m^2 modulus of rigidity G = 1×10^{11} N/m^2 density ρ = 7700 kg/m^3	0.03
barium titanate electrical capacitor	$k_dV_m^2/2\rho_d$	absolute permittivity k = 18×10^{-9} farad/m breakdown voltage $V_m = 5 \times 10^6$ V/m dielectric density ρ_d = 6000 kg/m^3	0.03
electrical inductance coil (non cored)	$B_m^2/2\mu\rho_c$	maximum flux density B_m = 2 volt.s/m^2 permeability $\mu = 2.5 \times 10^{-4}$ volt.s/amp.m core density ρ_c = 7700 kg/m^3	0.0001

flywheel and generator-motor would operate in an atmosphere of hydrogen or helium to reduce windage losses. Flywheels are not as yet a practical proposition mainly owing to the difficulties of designing bearings to operate satisfactorily at the necessarily high speeds which cause expansion and hence create large unbalance forces.

Powering vehicles with flywheels has been tried with some success, particularly in Switzerland and Germany, where buses were developed which ran on flywheel power between stops, recharging the flywheel with an electric motor at each stop by bringing an overhead trolley into contact with a power line. This system also took advantage of regenerative braking in which the motor acted as a generator which could be run during downhill driving, thereby charging a subsidiary energy store, namely a battery. This battery could in turn be used to supply lighting systems and other accessories. Regenerative braking has also been used in hydrostatic transmissions in which a hydraulic accumulator has been used as the store.

It should be noted that all energy stores are subject to losses of some sort or another, for example friction and windage losses in flywheels, leakage losses from pressure vessels, capacitors and thermal stores, and resistance losses in the coils of inductances. In most cases these losses are a small proportion of the rate of energy storage and extraction except for electrical inductances which require special attention to reduce such losses to a minimum.

Worked Example A 0.15 kW electric motor drives a punch crank which requires 200 joules of energy for each of its operating periods lasting 1 s at certain time intervals. The system is fitted with an energy store and at the start of each operating period the speed of the crank is to be 5 rev/s. How many operations can be performed per hour?

If the speed of the motor must not drop to less than 4.75 rev/s, deduce the kind of energy store required and determine its value. Assume that for this small speed fluctuation the power output of the motor is independent of speed.

Solution Let us first consider the storage elements at our disposal. Since the system is electrical-mechanical the storage elements worth considering are the inductance, the capacitance, the flywheel and the torsion spring. The first two would be installed in the electrical transmission between the supply and the motor and the second two in the mechanical transmission between the motor and the punch. In order to ensure the smoothest running of a system it is reasonable to instal any energy store as near as possible to the load whose demand is fluctuating. Hence in this example it is better to consider the mechanical stores of the flywheel or torsion spring, of which the former is the obvious choice since its mass effectiveness is by far the higher.

Consider now the number of operations possible per hour. n operations per hour demand 200n joules of energy. But the energy delivered by the motor in one hour = 150 x 60 x 60 joules.

Thus $$150 \times 3600/200 = 2700/\text{h}.$$

Now assume that the energy store is a flywheel of moment of inertia J. Thus the energy provided by the motor during 1 s is 150 joules. But the energy given up by the flywheel during 1 s

$$= \tfrac{1}{2}J\,(5^2 - 4.75^2)\,4\pi^2 \text{ joules},$$

while the energy demand by the machine during 1 s is 200 joules.

Thus $$150 + \tfrac{1}{2}J\,(5^2 - 4.75^2)\,4\pi^2 = 200,$$

whence $$J = 1 \text{ kg m}^2.$$

A similar problem is met in the design of the transmission systems of battery-operated vehicles. Here the power to the drive motor is varied by altering the spacing between power pulses from the battery by switching, but maintaining a constant battery output voltage level, which defines the height of the pulses. Attention must be paid to the electrical constants of the motor as well as to the mechanical properties of the transmission system to ensure adequately smooth operation.

Worked Example A rigid cylinder 10 m long and 2.5 m inside diameter is used in a chemical plant and, during the operation of the chemical process, the pressure of the fluid inside varies from a uniform 1000 kN/m^2 to a uniform 5000 kN/m^2. The fluid has a bulk modulus of 1×10^6 kN/m^2.

Deduce (a) the value of the 'fluid capacitance' of the cylinder in $m^3/(N/m^2)$. (b) the energy in Joules stored as a result of the increase in pressure.

Solution (a) From a comparison of equations (4.5.5) and (4.5.9) the fluid capacitance may be expressed as

$$C = \nu/K_B \ m^3/(N/m^2).$$

Volume $\nu = 10 \times \pi \ . \ 2.5^2/4 = 49 \ m^3.$

Hence $C = 49/(1 \times 10^6 \times 1000) = 49 \times 10^{-9} \ m^3/(N/m^2).$

(b) Energy stored $= \frac{1}{2}C \ . \ P^2 = \frac{1}{2} \times 49 \times 10^{-9} \ |5^2 - 1^2| \ 10^6 \times 10^6$

$$= 590 \ kJ.$$

Worked Example A storage tank is of inverted conical shape having a cone angle of 2 $\tan^{-1}$ 0.5 and a maximum depth equal to its maximum diameter. The pressure at the base is P.

(a) Deduce an expression for the stored energy in terms of P.

(b) Deduce an expression for its capacitance in $m^3/(N/m^2)$ in terms of P.

Solution (a) From the integral leading to equation (4.5.8)

Stored energy
$$\varepsilon_h = \int P \ . \ \frac{A}{\rho g} \ dP$$
$$= \frac{1}{\rho g} \int AP \ . \ dP.$$

Now, for any height H and diameter D,

$$P = \rho g H \text{ and } H = D.$$

Hence $A = \pi D^2/4 = \pi H^2/4 = (\pi/4) \ . \ (P/\rho g)^2$

and
$$\varepsilon_h = \frac{1}{\rho g} \int \frac{\pi}{4} \ . \ \left(\frac{P}{\rho g}\right)^2 \ P \ dP$$
$$= \frac{\pi}{4} \ . \ \frac{1}{(\rho g)^3} \int P^3 \ dP$$

$$= \frac{\pi P^4}{16\rho^3 g^3}.$$

(b) Comparing equations (4.5.8) and (4.5.9),

capacitance $$C = \frac{A}{\rho g} = \frac{\pi}{4} \cdot \left(\frac{P}{\rho g}\right)^2 \frac{1}{\rho g} = \frac{\pi P^2}{4\rho^3 g^3}\ \mathrm{m^5/N}.$$

4.7 Time Fluctuations of Converter and Load Parameters

As has been seen, one of the main functions of a storage element is to smooth out fluctuations in the potential or in the rate parameter which would otherwise occur if the element were not present. Let us now attempt to deduce, for given fluctuations in the converter output and in the load demand, the resulting time function of the rate parameter say when a rate-energy store is incorporated.

Consider again the flywheel of moment of inertia J storing rate (or kinetic) energy and acting as a smoother between a converter and a load. The converter output torque is considered to be fluctuating owing to variations in energy supply, say, or owing to the need to accelerate and decelerate masses and links in the converter (perhaps the piston and connecting rod of an internal combustion engine). The demand torque is fluctuating, say owing to resistance variations or owing to the need to accelerate and decelerate masses and links in the load device.

We may apply Newton's second law and write

$$T_c\ (N, \dot{N}, t) - T_L\ (N, \dot{N}, t) = J\dot{N}, \tag{4.7.1}$$

where J is the inertia referred to the converter shaft and T_c and T_L are the converter output-torque and the referred load demand-torque respectively, both being functions of converter speed N, acceleration $\dot{N}$ and time t. In general these equations will be difficult to solve. However, if we remember that the purpose of the storage element J is to ensure that fluctuations in speed are *small*, then we may write,

$$T_c\ (N, t) - T_L\ (N, t) \simeq JN\ .\ dN/d\theta \simeq JN_o\ dN/d\theta. \tag{4.7.2}$$

This assumes that the linkage masses are small enough to neglect $\dot{N}$ terms in comparison with the other terms in $T_c\ (N, \dot{N}, t)$ and $T_L\ (N, \dot{N}, t)$ and therefore N may be regarded as sensibly constant. The fact that T_c and T_L are functions of time allows us to still take account of accelerations of any masses within the converter or within the load during their cycles of operation provided these accelerations may be assumed to have negligible effect on the converter speed or load speed.

If θ is the angle of rotation of the converter output shaft from some datum position it may thus be assumed to be proportional to time t, such that

$$\theta = N_o t.$$

Thus, we may replace t by θ in equation (4.7.2) and integrate, giving

$$\int \left| T_c\ (N_o, \theta) - T_L\ (N_o, \theta) \right| d\theta = JN_o N$$

or

$$\int T\ (N_o, \theta)\ d\theta = JN_o N, \tag{4.7.3}$$

where $$T(N_o, \theta) = T_c(N_o, \theta) - T_L(N_o, \theta).$$

In most cases $T_c(N_o, \theta)$ and $T_L(N_o, \theta)$ may be written as trigonometric series which can usually be plotted in a similar way to that shown in Fig. 4.7a. Maximum speeds occur at points such as A since they each follow a period of excess converter output torque. Conversely points such as B follow periods of excess demand torque and therefore correspond to minimum speeds. By inspection, points of absolute maximum speed and absolute minimum speed may usually be identified. Let the corresponding values of θ be θ_1 and θ_2 respectively. Using equation (4.7.3) we may then write

$$\int_{\theta_2}^{\theta_1} T(N_o, \theta)\, d\theta = JN_o (N_{max} - N_{min})$$

$$= JN_o \Delta N_o, \text{ say,}$$

where ΔN_o is termed the maximum speed fluctuation. This is usually the most significant feature of the performance of a flywheel acting as an energy store. Summarizing then,

$$\Delta N_o = \frac{1}{JN_o} \int_{\theta_2}^{\theta_1} T(N_o, \theta)\, d\theta. \qquad (4.7.4)$$

<u>Worked Example</u> An engine drives directly an air compressor at a mean speed of 10 rev/s. The output torque T_c of the engine taking due account of the torque needed to accelerate its reciprocating parts is given by

$$T_c = 10{,}000 + 1{,}500 \sin\theta \text{ Nm},$$

while the demand torque T_L of the air-compressor is given by

$$T_L = 10{,}000 - 1{,}000 \sin 2\theta \text{ Nm},$$

where θ is the angle of rotation from some datum position (Fig. 4.7a). In order to ensure small fluctuations in speed, a flywheel of moment of inertia 1000 kgm^2 is used as a store, being interposed between the engine and the compressor.

Deduce the maximum speed fluctuations of the system and the mean power output of the engine in kW.

<u>Solution</u> Fig. 4.7a shows graphs of T_c and T_L on a base of θ.

Thus $$T(N_o, \theta) = T_c - T_L = 1{,}500 \sin\theta + 1{,}000 \sin 2\theta.$$

Since N is measured in rev/s, equation (4.7.3) becomes

$$4\pi^2 JN_o N = \int (1{,}500 \sin\theta + 1{,}000 \sin 2\theta)\, d\theta$$

and using equation (4.7.4), the maximum speed fluctuation ΔN_o is given by

$$\Delta N_o = \frac{1}{4\pi^2 JN_o} \int_{\theta_2}^{\theta_1} (1500 \sin\theta + 1000 \sin 2\theta)\, d\theta,$$

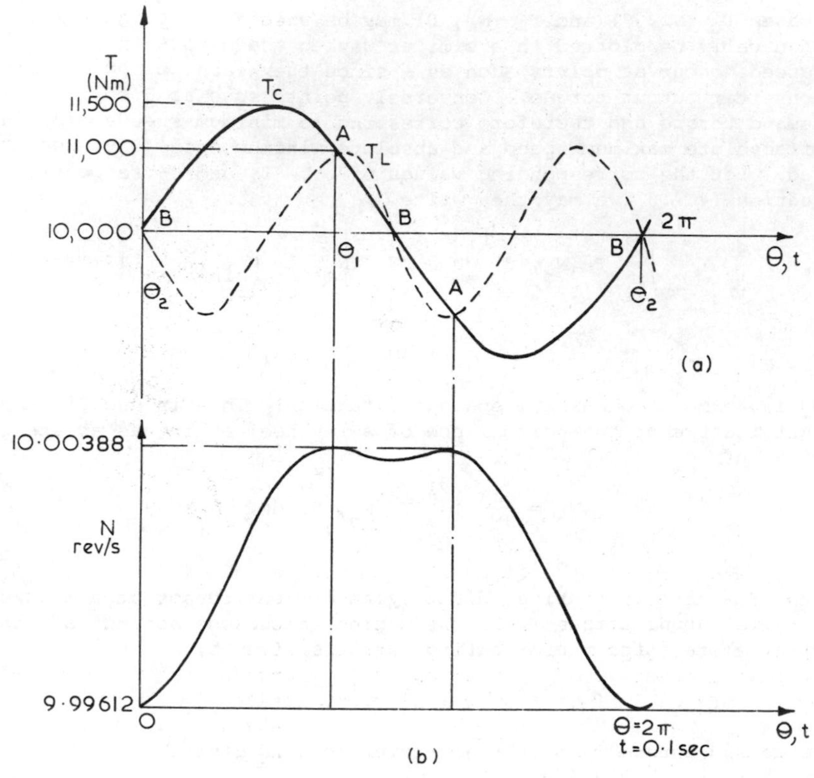

Fig. 4.7a, b. Torque and speed with flywheel storage element

in which θ_1 and θ_2 are found by equating T_C and T_L. In this case

$$1500 \sin \theta_{1,2} = - 1000 \sin 2\theta_{1,2}$$

that is $\theta_1 = 138.5^{\circ}, 498.5^{\circ}$ etc.

and $\theta_2 = 0^{\circ}, 360^{\circ}$, etc.

These give

$$\Delta N_o = \frac{1}{4\pi^2 \times 1000 \times 10} \int_0^{138.5} (1500 \sin \theta + 1000 \sin 2\theta)\, d\theta$$

$$= 0.00776 \text{ rev/s.}$$

If required a graph may be plotted of N versus θ by using equation (4.7.3). This takes the form shown in Fig. 4.7b.

Finally, the mean power output

$$= \text{mean torque} \times \text{mean speed}$$

$$= 10{,}000 \times 2\pi \times 10 = 628 \text{ kW}.$$

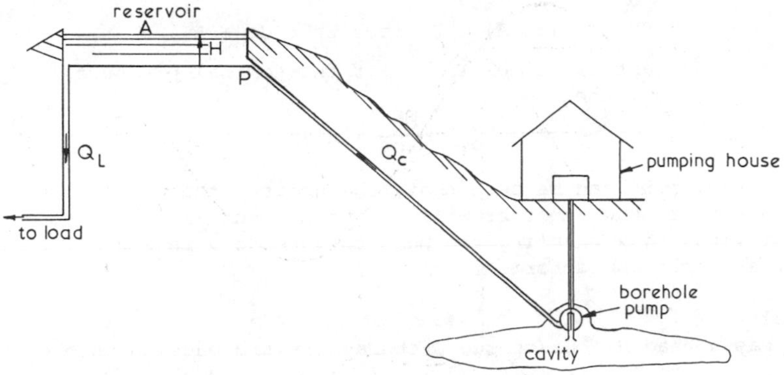

Fig. 4.7c. Use of a reservoir as a smoother

As a complementary study to that of the flywheel storing rate-energy let us now consider the use of a hydraulic potential-energy storage device in the system depicted in Fig. 4.7c. This is a schematic representation of a typical water-supply system used in the South Downs in which water is drawn from subterranean cavities and pumped to a reservoir, whence it is drawn off as required by the nearby inhabitants, which constitute the load. The purpose of the store is to enable the pump to deliver water against a constant pressure and hence at a constant rate even though the demand may be fluctuating considerably, particularly because of the difference between night-time and daytime usage.

We may thus write

$$Q_c - Q_L = A\dot{H},$$

where A is the cross-sectional area of the reservoir. If we neglect losses due to pipe friction, then the pressure P in the system at the base of the reservoir may be assumed to be given by

$$P = \rho g H,$$

where ρ is the fluid density.

Hence
$$Q_C - Q_L = (A/\rho g)\,\dot{P}. \tag{4.7.5}$$

Now assume that, owing to the presence of the store, any variations in pressure P are small. Hence the output flow Q_C of the pump may be assumed constant at Q_O and the demand flow Q_L may be assumed to be independent of P, just as in the previous study of the flywheel store, the demand torque was assumed to be independent of speed. Let us assume that the demand Q_L is given by

$$Q_L = Q_o + Q \cos 2\pi\Omega t,$$

in which frequency Ω is probably equivalent to about 3 cycles per day, that is

$$\Omega = 3/(24 \times 3600) \text{ Hz}.$$

Substitution into equation (4.7.5) gives

$$Q_o - (Q_o + Q \cos 2\pi\Omega t) = \dot{P}A/\rho g$$

and hence
$$P = -(\rho g/A) \,.\, Q \int \cos 2\pi\Omega t \, dt + A_1,$$

where A_1 is a constant of integration. If the nominal pressure is P_o and $P = P_o$ at $t = 0$, say, then

$$P = P_o - \frac{\rho g Q}{2\pi\Omega A} \sin 2\pi\Omega t.$$

From this equation it can be seen that, the greater the cross-sectional area A of the reservoir, the smoother will be the pressure in the system as a function of time, thus enabling the pump to work at a reasonably constant point on its output characteristic.

4.8 Examples

1. A pump has a head H - discharge Q characteristic passing through the following (Q,H) points (0 litres/s, 33 m), (10,300), (20,25), (30,20), (40,10) and (46,0). It lifts water from a lower tank having no inlet to an upper tank having no outlet. The tank dimensions are 10 m x 5 m in horizontal cross-section. Pumping begins when the upper tank is empty and the difference in water level is 13 m. The total quantity of water to be lifted is 200,000 litres. Find the time taken. Neglect pipe friction losses.

   (Ans. Approx. 100 minutes).

2. A 30 hp 60 Hz induction motor has a speed-torque characteristic passing through the following points:

Speed (% of synchronous speed)	100	95	50	0
Torque (% of rated torque)	0	300	150	150

   The motor and its direct-coupled load possess a total moment of inertia of 50 kg m^2.

   When a constant resistive torque of 120 Nm is being overcome the circuit-breaker opens and the motor speed falls. How long can the breaker remain open for the motor to regain its previous speed on reapplication of the power? (1 HP = 746 watts).

   (Ans. 37.5 s).

3. Show that, for a d.c. series motor the output torque-speed characteristic may be classed in category (iv) as discussed in Chapter 2. Why is its characteristic so suitable for applications in which large masses have to be accelerated (as in lifts, cranes and traction) and what precautions must be taken in the use of such a converter?

4. The torque T delivered by a motor can be expressed over the working range by

$$T = 2000/N \text{ Nm},$$

where N is the speed in rev/s. The driven load is a direct-coupled fan whose torque demand is given by

$$T = N^2/32 \text{ Nm.}$$

Find graphically the time taken for the speed to rise from 20 to 35 rev/s, if the moment of inertia of the fan together with the other rotating components is 4 kg m^2.

Plot the speed-time curve between 20 and 35 rev/s. Write down a differential equation relating T, N and time.

(Ans. Approx. 8 s.)

5. A 50 Hz induction motor drives a lubricating-oil pump through a gearbox for which output speed/input speed = 1.5. The inertia of the rotating parts of the pump is 1 kg m^2 and the torque demand at the pump is given by

$$T_p = 0.4\ N_p \text{ Nm,}$$

where N_p is the pump speed in rev/s. The output torque of the motor is given by

$$T_m = 20 + 0.9\ N_m \quad (45 > N_m \geqslant 0)$$

and $$T_m = -\ 12\ N_m + 600 \quad (50 \geqslant N_m > 45),$$

where N_m is the motor speed in rev/s.

Neglecting the moments of inertia of the gear wheels, find the operating speed and the time taken to reach a speed of 45 rev/s from rest. Work from first principles throughout (that is, assume only Newton's laws of motion).

(Ans. 46.5 rev/s, 31.8 s.)

6. A gear pump is driven by a 50 Hz induction motor through a 2:1 step-down gearbox. Without leakage it would discharge 0.05 litres of oil per revolution. The leakage path through the pump clearances is equivalent to a feedback path across the pump. The leakage flow is proportional to the pressure developed by the pump and is 30 litres/min when the pressure is 350 kN/m^2.

The torque-speed demand characteristic of the pump is given by

$$T = 0.1\ N \text{ Nm where N is measured in rev/s}$$

and the torque-speed output characteristic of the motor is given by

$$T = 3.875 + 0.025\ N \text{ Nm for N between 0 and 45 rev/s}$$

and $$T = -\ N + 50 \text{ Nm for N between 45 and 50 rev/s.}$$

The moments of inertia of the rotating parts attached to the motor and pump shafts are 6×10^{-2} and 16×10^{-2} kg m^2 respectively.

(a) Find the operating speed of the pump and the time taken to reach 45 rev/s from rest.

(b) The pump lifts oil of density 900 kg/m^3 from a lower tank having no inlet to an upper tank having no outlet. The tank dimensions are 1 m x 1 m in horizontal cross-section. Pumping begins when the upper tank is empty and the difference in oil level is 13 m. The total quantity of oil to be lifted is 4000 litres. Find the time taken assuming the operating speed found in (a) and neglecting pipe friction losses.

(Ans. 24.5 rev/s, 7.3s; 0.55h.)

7. A compressor fills an air bottle of volume v through a pipe. Assuming output and demand characteristics similar to those of Fig. 4.2b, evolve a graphical method of estimating the time taken to pressurise the air vessel from a pressure P_1 to a pressure P_2. Assume that the effective bulk modulus of the air in the pressure vessel at any time is equal to the instantaneous pressure in the vessel.

(Ans. $\Delta\ (P_C - P_L)/Q = (P_C - P_L)\ \Delta t/v$).

8. A pump has a pressure-discharge characteristic which may be represented by the equation

$$P = 300 - 7Q \text{ kN/m}^2,$$

where Q is the discharge in litres/s. It pumps water into a reservoir of constant surface area 50 m^2 and the demand characteristic seen by the pump is given by

$$P = 9.81\ H + 5Q \text{ kN/m}^2,$$

where H is the level of water in metres in the reservoir and Q is the flow into the reservoir in litres/s.

Determine graphically or otherwise the time taken to fill the reservoir from a level of 10 m to a level of 20 m.

(Take $g = 9.81\ m/s^2$, $\rho = 1000\ kg/m^3$.)

(Ans. 11 h.)

9. A motor vehicle, total mass 1360 kg has road wheels of 0.63 m effective diameter. The effective moment of inertia of all four road wheels and of the driving axle together is 6.7 kg m^2, while that of the engine and flywheel is 0.84 kg m^2. The tractive resistance at a speed of 24 km/h is 266 N and the total available engine torque is 203 N m.

What would be the required gear ratio, rear axle speed to engine speed, to provide maximum possible acceleration on an up-grade whose sine is 0.25, when travelling at 24 km/h?

(Ans. 1/19).

10. An epicyclic gear train consists of a sun wheel B having 36 teeth and a polar moment of inertia 52.5×10^{-4} kg m^2; 3 planet wheels P each having 12 teeth, a polar moment of inertia 9×10^{-4} kg m^2 and mass 0.91 kg; a spider A having a 0.076 m arm and a polar moment of inertia 0.084 kg m^2 and a fixed outer annulus C having 60 teeth.

A torque is applied at the input shaft to the sun wheel and at a given instant the sun wheel has an angular velocity of 15 rad/s and an angular acceleration of 60 rad/s^2. There is a resisting torque of 0.735 N m acting on the output shaft connected to the spider.

Find the value of the applied torque at this instant and the velocity of rubbing at the planet wheel pin which is 0.013 m diameter.

(Ans. 1.8 N m, 182 mm/s.)

11. The effective turning-moment exerted at the crankshaft of an engine is represented by

$$T = 12200 + 1525 \sin 2\theta - 3050 \cos 2\theta \text{ N m},$$

where θ = inclination of the crank from top dead centre. The mass of the flywheel is 1710 kg and its radius of gyration 0.76 m. The mean engine speed is 2 rev/s and the external resistance is constant.

Find the maximum speed fluctuation and the power developed by the engine.

(Ans. 0.0435 rev/s; 153 kW).

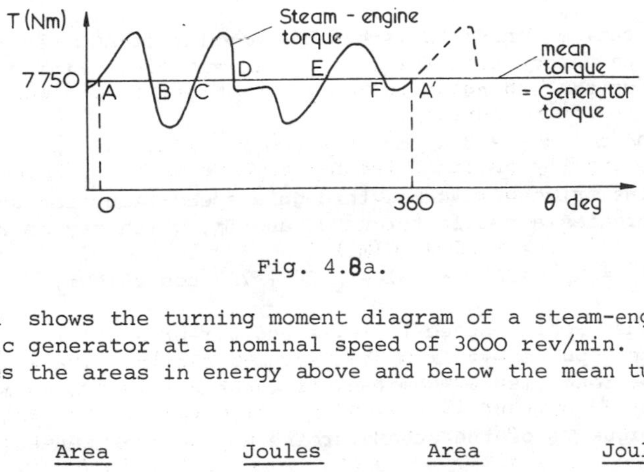

Fig. 4.8a.

12. Fig. 4.8 a shows the turning moment diagram of a steam-engine which drives an electric generator at a nominal speed of 3000 rev/min. The following table gives the areas in energy above and below the mean turning moment line:

Area	Joules	Area	Joules
AB	1400	BC	1900
CD	1400	DE	1800
EF	930	FA'	30

The moment of inertia of the flywheel on this engine is 200 kg m^2. Assuming speed fluctuations from the mean speed are small, and that the demand torque from the generator is constant, calculate the maximum speed fluctuation about the mean.

(Ans. $\pm$ 0.352 rev/min.)

13. A workshop compressor supplies air to a fluctuating demand and the supply line is fitted with a pressure vessel to smooth out variations in pressure in the system. The nominal system pressure is P_o and temperature variations may be assumed negligible. The following parameters relate to the system:

Volume of system	ν
Output flow from compressor	Q_o
Demand flow	$Q_o + Q \cos \omega t$

Deduce an expression for system pressure P as a function of time, if it is assumed that $P = P_o$ when $t = 0$.

(Ans. $P = P_o\ (1 - (Q/\omega\nu) \sin \omega t)$.)

14. An engine provides an output torque T_E which may be approximated by

$$T_E = 15000 + 4665 \sin 3\theta \text{ Nm},$$

while the demand torque T_L of its direct-coupled load may be written

$$T_L = 15000 + 2000 \sin \theta \text{ Nm}.$$

In these equations θ is the output shaft angular displacement from some datum position. The total moment of inertia J on the connecting shaft is given by

$$J = 1000 \text{ kg m}^2.$$

Assuming the engine (and load) speed has a value of 600 rev/min when $\theta = \pi/2$, deduce an equation for speed in terms of θ. Find also the mean output power of the engine.

(Ans. $N = 3.04 \cos \theta - 2.364 \cos 3\theta + 600$ rev/min, 942 kW.)

15. A steam turbine drives a machine through a speed-reduction gear of 30:1. The machine creates a cyclic torque demand T_L, which may be represented by

$$T_L = N_L^2\ |3050 - 381 \sin 2\theta + 763 \cos 2\theta|\ \text{Nm},$$

where N_L is the machine speed in rev/s and θ represents the angular position of the machine input shaft from some datum position. The moment of inertia of the turbine rotor is 10 kg m^2 and of the machine 1000 kg m^2.

The output torque T_c of the steam turbine may be represented by

$$T_c = 500 - 0.026\ N_c^{\ 2} \text{ Nm},$$

where N_c is the turbine speed in rev/s.
Under equilibrium conditions the mean machine speed is 2 rev/s and occurs when $\theta = 0, \pi\ , 2\pi\ , \ldots.$ etc.

(a) Deduce an equation relating machine speed and time, assuming speed fluctuations to be small compared with the mean speed.

(b) Estimate roughly the time taken for the turbine to accelerate the machine up to a speed of 1 rev/s from rest. A graphical method may be employed and cyclical torque components may be disregarded for this part of the question.

(Ans. (a) $N_L = 2 + 9.7 \times 10^{-4} (1 - \cos 2\theta) - 19.3 \times 10^{-4} \sin 2\theta$ rev/s.
(b) Approx. 4.4 s.)

16. An engine has a torque-speed output characteristic given by

$$T_{oE} = 3N_{oE} - 0.02N_{oE}^{2} \text{ Nm.}$$

It drives a load whose torque-speed demand characteristic is given by

$$T_{iL} = N_{iL} \text{ Nm}$$

through a fluid coupling whose input (demand) torque and output torque are each given by

$$T_{i,o} = 0.06N_i^{2} \left[1 - (N_o/N_i)^4\right] \text{ Nm.}$$

Speeds N are in rev/s throughout.
The moments of inertia attached to the engine and load shafts are given by

$$J_E = J_L = 1 \text{ kgm}^2.$$

Evolve a calculation scheme for obtaining the engine speed from 1 rev/s and the load speed from zero with time as the latter is increased in specified increments Δt. Deduce the slip under equilibrium conditions.

(Ans. calculation scheme Fig. 4.8b; approx. 0.04.)

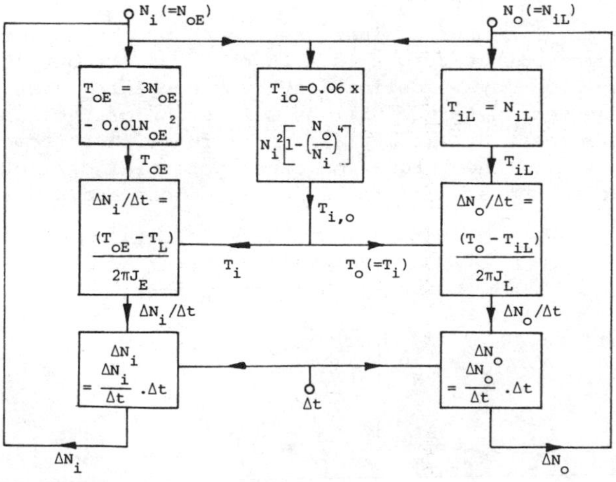

Fig. 4.8b. Calculation scheme

CHAPTER 5

THE LINEAR DYNAMICS OF MECHANICAL SYSTEMS

In the last chapter the performance of a system in acquiring and maintaining its steady-state operating condition was discussed. Here we shall consider the system performance in acquiring a *new* steady state operating condition as a result of a *change* in output or demand characteristic. For this purpose we shall assume that the system dynamics may be considered on the basis of linear theory as this makes the mathematics easy to deal with.

5.1 Small Excursion Dynamics

Consider a system whose load is represented by the torque-speed demand characteristic (referred to the converter) shown as curve *a* in Fig. 5.1a. If the referred load torque suddenly increases owing to the sudden introduction of extra friction in the load, say, the effect on the load characteristic might be to reposition it from curve *a* to curve *b*. This alters the matching point from A to B on the converter output characteristic x_1y_1 and results in a decrease in converter speed $(N_0 - N_{01})$. Alternatively, it is possible for the converter output characteristic to change suddenly to x_2y_2, say, owing to a disturbance in its power or fuel supply and an increase in converter speed $N_{02} - N_0$ might result. To be rather more general let us assume that alterations occur to both the converter and to the referred load characteristics. As a result let the *increase* in converter output torque be ΔT_C and the *increase* in load demand torque as seen by the converter be ΔT_L. As we have seen in section 4.4 we may write

$$\Delta T_C - \Delta T_L = J\ \delta\ (\Delta N)/\delta t \qquad (5.1.1)$$

where J is the total moment of inertia as seen by the converter. Now we know that the converter output torque is a function of its output speed N and of its supply of fuel which may be defined say by the opening β of its fuel supply valve. Also the demand torque will be a function of the load speed and hence of the converter speed N. It may also be a function of some other resistance parameter, say R. If the torque (or other potential parameter) functions are linear as may often be assumed for electric motors and generators then we may write

$$\Delta T_C = \left.\frac{\partial T_C}{\partial N}\right|_\beta \Delta N + \left.\frac{\partial T_C}{\partial \beta}\right|_N \Delta\beta = C_N\Delta N + C_\beta\Delta\beta\text{, say}$$

$$(5.1.2)$$

and

$$\Delta T_L = \left.\frac{\partial T_L}{\partial N}\right|_R \Delta N + \left.\frac{\partial T_L}{\partial R}\right|_N \Delta R = L_N\Delta N + L_R\Delta R\text{, say}$$

where

$$C_N = \left.\frac{\partial T_C}{\partial N}\right|_\beta,\ C_\beta = \left.\frac{\partial T_C}{\partial \beta}\right|_N,\ L_N = \left.\frac{\partial T_L}{\partial N}\right|_R \text{ and } L_R = \left.\frac{\partial T_L}{\partial R}\right|_N.$$

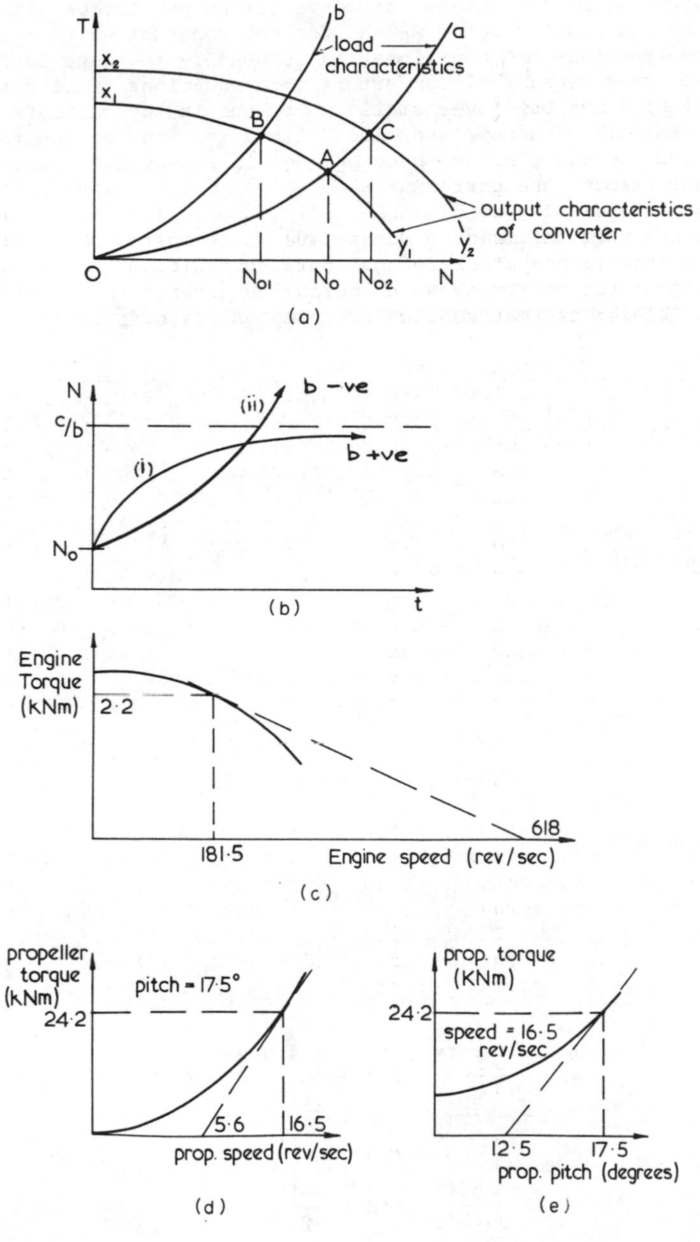

Fig. 5.1a - e. Torque characteristics

ΔN, $\Delta\beta$ and ΔR represent the *increases* in the *independent parameters* N, β and R respectively from their equilibrium values N_0, β_0 and R_0 and $\partial T_C/\partial N$, for example, denotes the rate of change of converter output torque with converter speed about the equilibrium speed N_0 and for the constant valve opening, β_0. If however the functions are non-linear as is usually the case for engines, pumps and most other mechanical converters then equations similar to equations (5.1.2) still apply but only over small excursions in the parameters. The partial differentials then represent local linearisations of the torque characteristics around the equilibrium speed N_0, equilibrium valve opening β_0 and equilibrium resistance parameter R_0.

The use of the term *independent parameters* may require some elucidation. If we consider for example a centrifugal pump driven from a mechanical power source the pump has inputs T and N and outputs P and Q. The important internal feature of the black-box representing the pump is its efficiency η given by

$$\eta = \frac{PQ}{TN}.$$

We might be tempted to write

$$P = f\ (T,N,Q)$$

and differentiate accordingly. However if we realise that $Q = \eta TN/P$, then it becomes obvious that for a given pump,

$$P = f\ (T,N,\frac{\eta TN}{P})$$

or

$$P = f\ (T,N)$$

and accordingly

$$\Delta P = \left.\frac{\partial P}{\partial T}\right|_N \cdot \Delta T + \left.\frac{\partial P}{\partial N}\right|_T \cdot \Delta N.$$

Alternatively we may write

$$P = f\ (T,Q)$$

or

$$P = f\ (N,Q),$$

provided only three performance parameters enter the equation.

Upon differentiation the latter two equations give

$$\Delta P = \left.\frac{\partial P}{\partial T}\right|_Q \cdot \Delta T + \left.\frac{\partial P}{\partial Q}\right|_T \cdot \Delta Q$$

and

$$\Delta P = \left.\frac{\partial P}{\partial N}\right|_Q \cdot \Delta N + \left.\frac{\partial P}{\partial Q}\right|_N \cdot \Delta Q$$

respectively.

Returning to the problem in hand, substitution of equations (5.1.2) into equation (5.1.1) gives

$$(C_N \Delta N + C_\beta \Delta_\beta) - (L_N \Delta N + L_R \Delta R) = J \frac{\delta}{\delta t} (\Delta N). \qquad (5.1.3)$$

Since ΔN, $\Delta\beta$ and ΔR refer to *increases* from initial equilibrium values N_0, β_0 and R_0 we can now write

$$\Delta N = N - N_0$$

$$\Delta\beta = \beta - \beta_0$$

$$\Delta R = R - R_0,$$

and equation (5.1.3) becomes

$$C_N (N - N_0) + C_\beta (\beta - \beta_0) - L_N (N - N_0) - L_R (R - R_0) = J \frac{\delta}{\delta t} (N - N_0)$$

$$= J \frac{\delta N}{\delta t},$$

or
$$J \frac{\delta N}{\delta t} + N (L_N - C_N) = N_0 (L_N - C_N) + C_\beta (\beta - \beta_0) - L_R (R - R_0). \qquad (5.1.4)$$

We have seen from section 4.4 that, for stability, we require that

$$\frac{\partial T_C}{\partial N} < \frac{\partial T_L}{\partial N}.$$

In the notation of this section the stability requirement becomes

$$C_N < L_N$$

or
$$L_N - C_N > 1,$$

where $L_N - C_N$ can be seen to be the term in N in equation (5.1.4). For simplicity we may rewrite equation (5.1.4) in the form,

$$J \frac{\delta N}{\delta t} + bN = C,$$

where b and C are constants.

Its solution can be effected by integrating to give

$$\int dt = \frac{J}{b} \int \frac{b dN}{(C - bN)}$$

or
$$t = \frac{-J}{b} \ln (C - bN) + A,$$

where A is a constant of integration.

Thus
$$N = (1/b) \left| C - \exp (A - t) b/J \right|. \qquad (5.1.5)$$

Now when $t = 0$, $N = N_0$, the initial equilibrium speed and hence

$$N_0 = (1/b)\ |C - \exp Ab/J|.$$

Substituting the resulting expression for A into equation (5.1.5) gives

$$N = (1/b)\ |C - (C - bN_0)\exp(-\ tb/J)|. \qquad (5.1.6)$$

Equation (5.1.6) describes an increase in speed whose form is shown in Fig. 5.1b, curve (i). The final steady speed may be found by putting t equal to infinity in equation (5.1.6) or by putting $\delta N/\delta t$ equal to zero in equation (5.1.4). In either case, the final steady speed N_0' is given by

$$N_0' = C/b = N_0 + \frac{C_\beta\ (\beta - \beta_0)}{(L_N - C_N)} - \frac{L_R\ (R - R_0)}{(L_N - C_N)}. \qquad (5.1.7)$$

Consider now what would happen if b were negative (say $b = -B$). This means that $L_N - C_N < 1$, suggesting instability. Substitution into equation (5.1.6) now gives

$$N = -\ (1/B)\ |C - (C + BN_0)\exp tB/J|,$$

whose graph is shown as curve (ii) in Fig. 5.1b. This illustrates an exponential rise in speed which exemplifies an unstable state of affairs.

Worked example An engine drives a variable-pitch propeller at 16.5 rev/s through a speed-reducing gearbox of ratio 11:1. The engine torque-speed output characteristic at a constant throttle opening is shown in Fig. 5.1c and the propeller torque-speed demand characteristic for a pitch of 17.5° in Fig. 5.1d. The propeller torque-pitch characteristic at a propeller speed of 16.5 rev/s is shown in Fig. 5.1e. The moments of inertia of the engine rotating parts and of the propeller are 4.6 and 131 kgm^2 respectively.

If a small *reduction* of 0.2 degree occurs in the pitch setting of the propeller, formulate and solve the differential equation governing the resulting speed variation of the system, assuming that the engine throttle setting does not alter.

Solution It has been shown in section 4.2 that the inertia J 'seen' by the engine will be given by

$$J = 4.6 + 131/11^2 = 5.69\ kg\ m^2$$

and that the torque T_L 'seen' by the engine at the operating point will be given by

$$T_L = 24.2/11 = 2.2\ kN\ m.$$

Now

$$\Delta T_C - \Delta T_L = J\ \delta(\Delta N)/\delta t, \qquad (5.1.8)$$

where N is the engine speed in rad/s which under steady-state conditions is equal to 11 x 16.5 = 181.5 rev/s = 1140 rad/s. The only external disturbance occurs at the propeller. This will produce a speed change and hence changes in converter output torque and load demand torque as follows:

$$\Delta T_C = \frac{\partial T_C}{\partial N}\ .\ \Delta N$$

and
$$\Delta T_L = \frac{\partial T_L}{\partial N}\bigg|_p \Delta N + \frac{\partial T_L}{\partial p}\bigg|_N \Delta p, \tag{5.1.9}$$

where p denotes the pitch condition of 17.5°.

Now from Fig. 5.1c

$$\frac{\partial T_C}{\partial N} = \frac{-2.2 \times 10^3}{2\pi(618 - 181.5)} \text{(N m s)} = -0.8 \text{ N m s.} \tag{5.1.10}$$

In deducing $\frac{\partial T_L}{\partial N}\Big|_p$ and $\frac{\partial T_L}{\partial p}\Big|_N$ it must be remembered that T_L is the load torque referred to the converter and that N is the converter speed. This means that, to use Figs. 5.1d and 5.1e, we must divide all torque values by the gear ratio, namely 11. We must also multiply the propeller speed in Fig. 5.1d by 11.

Hence $$\frac{\partial T_L}{\partial N}\bigg|_p = \frac{\partial T_L}{\partial N}\bigg|_{17.5} = \frac{10^3 \times 24.2/11}{2\pi(16.5 - 5.6)11} = 2.86 \text{ N m s.}$$

Also $$\frac{\partial T_L}{\partial p}\bigg|_N = \frac{\partial T_L}{\partial p}\bigg|_{1140} = \frac{10^3 \times 24.2/11}{(17.5 - 12.5)} = 443 \text{ N m/degree.}$$

Substitution into equation (5.1.9) gives

$$\Delta T_L = 2.86\ \Delta N + 443\ \Delta p \tag{5.1.11}$$

in which we are given that $\Delta p = -0.2$ degree.
From equations (5.1.8), (5.1.10) and (5.1.11) we have

$$-0.8\ \Delta N - (2.86\ \Delta N - 88.6) = 5.69\ \delta(\Delta N)/\delta t,$$

where ΔN is the *increase* in engine speed in rad/s.
As the constant converter speed before the small pitch disturbance was 181.5 rev/s or 1140 rad/s, then we may write

$$\Delta N = N - 1140$$

and hence $$\delta(\Delta N)/\delta t = \delta(N - 1140)/\delta t = \delta N/\delta t,$$

and the differential equation governing the speed of the system is as follows,

$$5.69\ \delta N/\delta t + 3.66\ N = 3.66\ N_o + 88.6,$$

in which $N_o = 1140$ rad/s. This is again of the form

$$J\frac{\delta N}{\delta t} + bN = C,$$

and, by comparison with equation (5.1.6), the solution is

$$N = (1/3.66)\ |(3.66\ N_o + 88.6) - 88.6 \exp(-3.66t/5.69)|,$$

in which $N_o = 1140$ (rad/s).

Thus $$N = 1140 + 24.4\ |1 - \exp(-0.643t)|,$$

giving a speed-time graph of the form (i) shown in Fig. 5.1b, in which

$$N_o = 1140 \text{ rad/s}$$

and

$$C/b = 1164.4 \text{ rad/s.}$$

5.2 Restoration of a Specified System Parameter

In the previous section we referred to Fig. 5.1a, which showed the change in speed arising from changes in demand characteristic and converter output characteristic. We shall here consider means of restoring the converter speed to its desired original value, N_o. To do this we shall attempt to achieve a different converter output characteristic x_2y_2, say (Fig. 5.1a) by perhaps increasing the valve opening β in the fuel line to the engine (Fig. 5.2a). This results in the establishment of a new matching point A' and the original

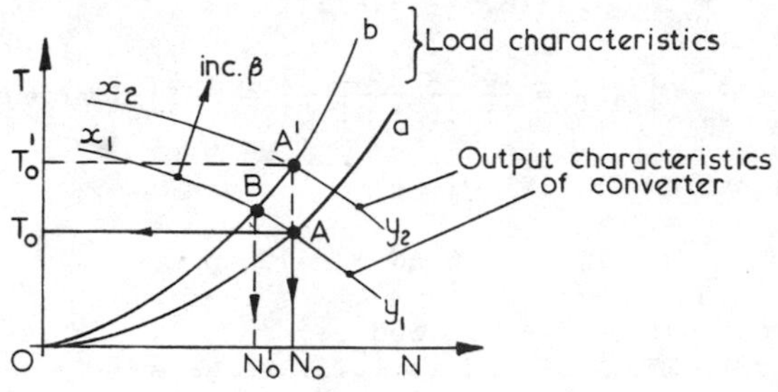

Fig. 5.2a. Restoration of speed by increasing throttle opening

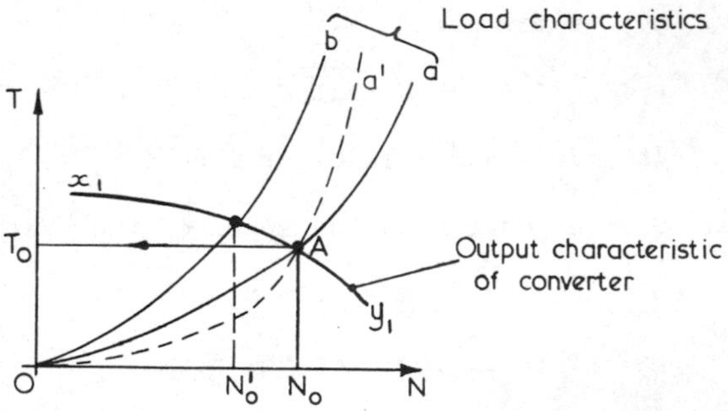

Fig. 5.2b. Restoration of speed by altering load parameter

speed N_o is restored, at the expense of an increased torque requirement T_o'. Alternatively we may resposition the load characteristic from curve *b* to curve *a'* (Fig. 5.2b) by the alteration of another load parameter to achieve the original matching speed N_o.

In a practical engineering system the first indication that the speed has decreased is usually obtained from a sensing device which is simply a small-scale energy conversion element or transmission element. The energy so obtained can be transmitted to another conversion or transmission element known as a control device (in this case the fuel valve) to carry out the remedial measure to restore the required condition. This is said to be employing feedback to achieve the remedy and a block diagram of a typical system complete with feedback path is shown in Fig. 5.2c. The combination of sensing device, comparator and transducer is known as a controller.

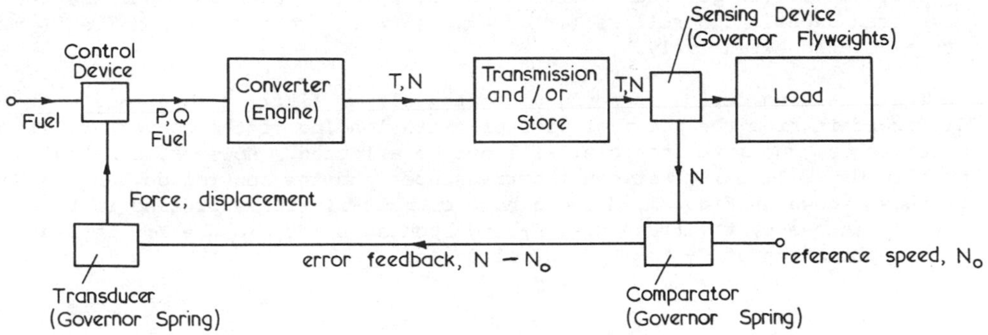

Fig. 5.2c. System block-diagram

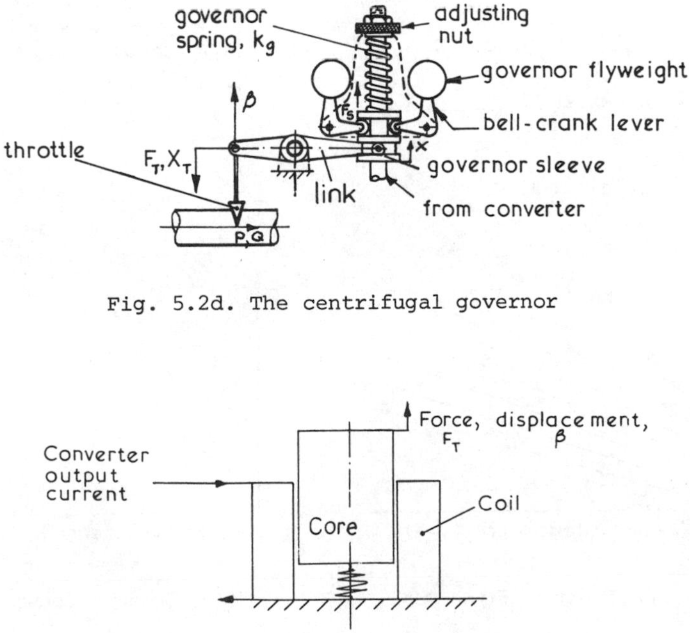

Fig. 5.2d. The centrifugal governor

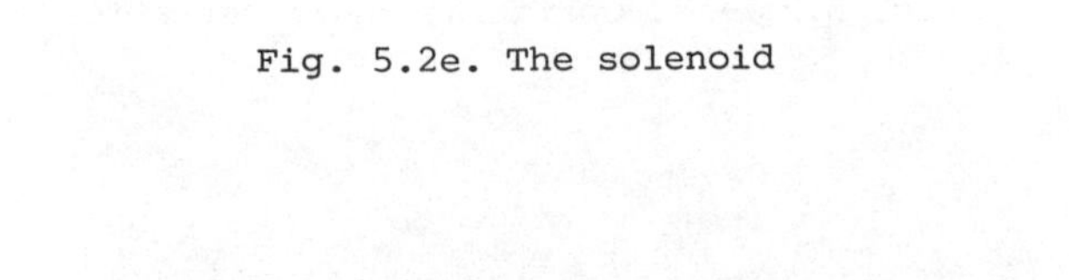

Fig. 5.2e. The solenoid

A good example of a controller in common use is the centrifugal governor (Fig. 5.2d) in which the governor flyweights constitute the sensing device and the governor spring constitutes the comparator plus transducer. The throttle, or in some cases a fuel pump, is the control device. The governor thus provides a force to change the throttle displacement in response to a change in speed at the converter output shaft. Another example of a controller is the solenoid (Fig. 5.2e) which provides a force and displacement in response to a change in current. Here the sensing device is the central iron core which operates in a magnetic field supplied by the current in the coil, just as the governor flyweights operate in a centrifugal field supplied by rotation of the governor shaft. The spring again acts as a comparator plus transducer. Controllers (or governors) with pressure-responsive elements are also used extensively.

The measuring device is designed to tap off so little torque that we may disregard it from the point of view of extra loading of the converter. In other words, the matching point will not be affected. However, a matching problem may well exist between the transducer and the control device. Both of those shown in Figs. 5.2d and e have characteristics expressed in terms of force produced by the transducer F_T and opening β (or closure X_T) obtained upon application of F_T to the control device.

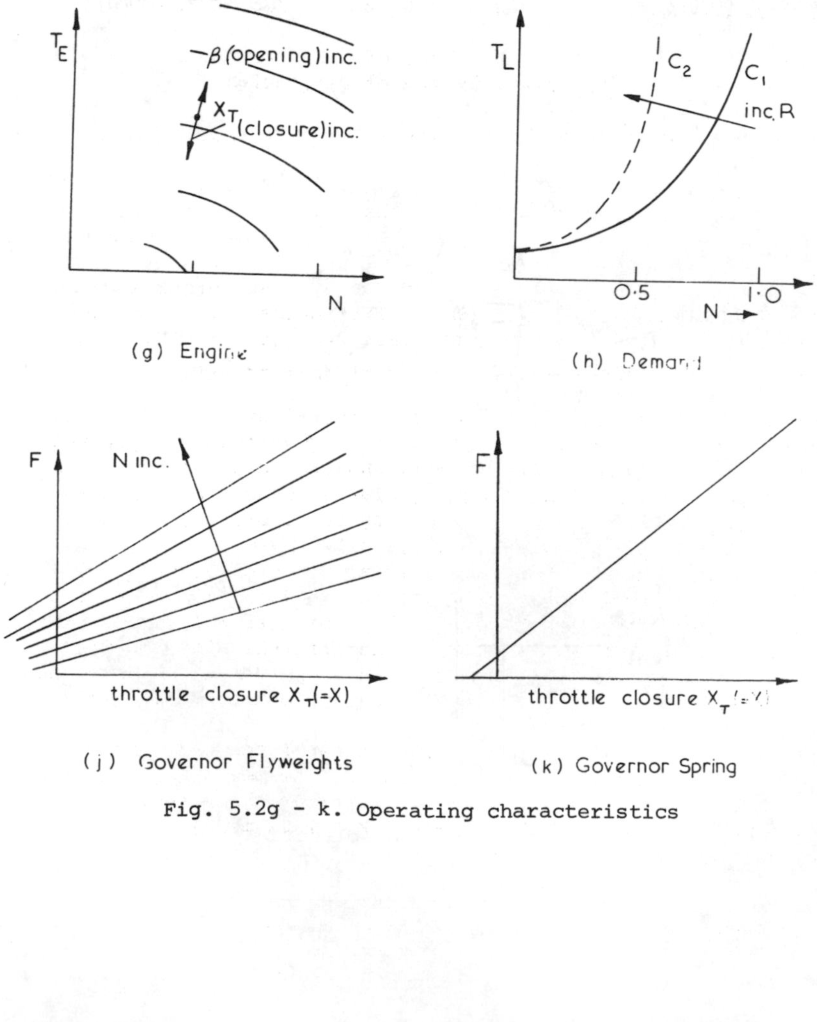

Fig. 5.2g - k. Operating characteristics

Worked example A reciprocating engine drives a load, the block diagram of the system being shown in Fig. 5.2c. A governor (Fig. 5.2d) governs the engine speed by providing a force F_T, and hence closure X_T to the engine fuel-valve or throttle in response to an increase in the speed N at the engine output shaft to which the governor is geared. Fig. 5.2c illustrates the feedback loop involving the governor. The torque speed output characteristics of the engine are given in Fig. 5.2g for different throttle-closures X_T, and the load-demand curve is given at C_1 in Fig. 5.2h. The F-X output characteristics of the speed-sensing portion of the governor (the flyweights) are given in Fig. 5.2j while Fig. 5.2k shows the force-deflection demand characteristic of the governor spring. Show how the operating point may be obtained in co-ordinates of speed N and throttle closure X_T. Discuss what happens if the load characteristic suddenly changes from C_1 to C_2 (Fig. 5.2h).

Solution Assuming the arms of the governor-throttle link in Fig. 5.2d are equal we may then assume that in the absence of friction $F_S = F_T$ and $X = X_T$. The procedure is then as follows:

(1) Combine the curves of Figs. 5.2j and k to give possible matching points X_1 to X_5 corresponding to different speeds N_1 to N_5 (Fig. 5.2l).

(2) Combine the curves of Figs. 5.2g and h to give different matching speeds N_A to N_E for different valve closures X_A to X_E (Fig. 5.2m).

(3) On axes of N and X_T plot the values of N_1 to N_5 against the corresponding values of X_T (X_1 to X_5) to give curve *a* of Fig. 5.2n.

(4) On the same axes plot N_A to N_E versus X_A to X_E and obtain curve *b* of Fig. 5.2n.

(5) The intersection of curves *a* and *b* then gives the required operating point O.

If N_o is the desired speed, X_o on the control valve is set by preloading the governor spring to F_o (Fig. 5.2l). This means that at a speed N_o the centrifugal force of the flyweights equals the total spring force when the spring deflection is such that the valve closure is X_o.

If the load characteristic suddenly changes from C_1 to C_2 (Fig. 5.2h) the same governor curves are used (Fig. 5.2l) to give the same governor N - X characteristic *a* as before (Fig. 5.2p). However a different engine characteristic *b'* is now obtained from C_2 using Fig. 5.2q and this results in a new operating point O'. From Fig. 5.2p it may be observed that an error in speed (N_o - N_o') exists. It is evident that, while the governor is having a beneficial effect in that, for the same initial valve opening X_o, it is not allowing the engine-speed to drop from N_o to N_o" (Fig. 5.2q), there is nevertheless some error in speed N_o - N_o' (Fig. 5.2p). This error is to be expected since, if the engine speed had been restored, the original valve opening would also have been restored by governor action. This would not then have been sufficient to allow enough fuel to cope with the greater load-torque.

It is worthwhile presenting the discussion of the last paragraph mathematically to give an alternative appreciation of the performance of the governor. From Fig. 5.2n assume that the slope of governor characteristic *a* at the operating point O is given by $\Delta N/\Delta X_T = 1/k$. This means that, in terms of throttle opening,

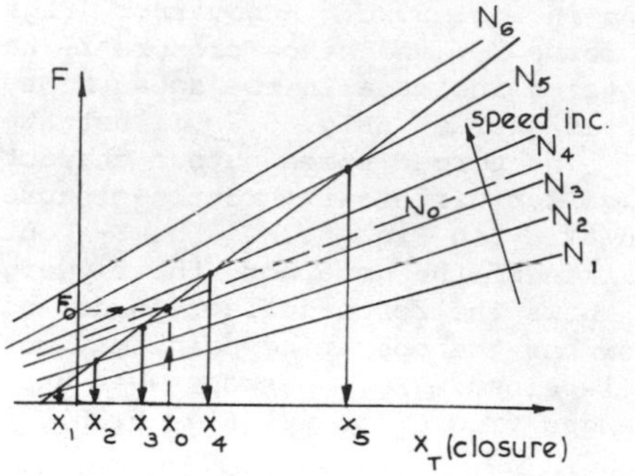

(l)

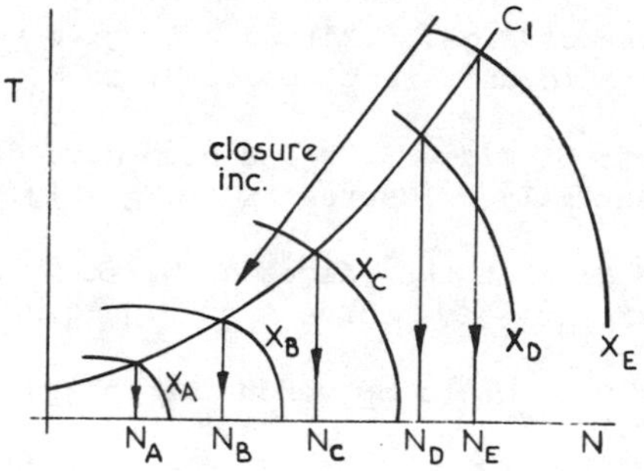

(m)

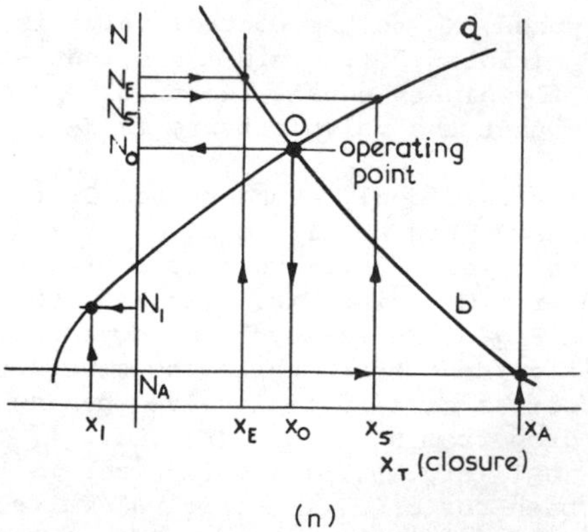

(n)

Fig. 5.2 – l – n solution procedure

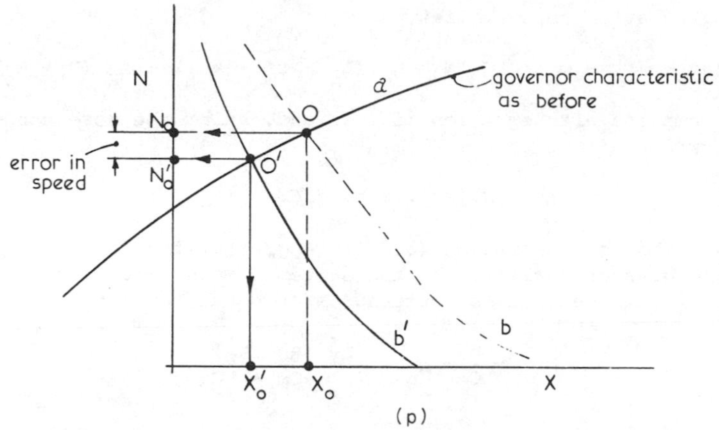

(p)

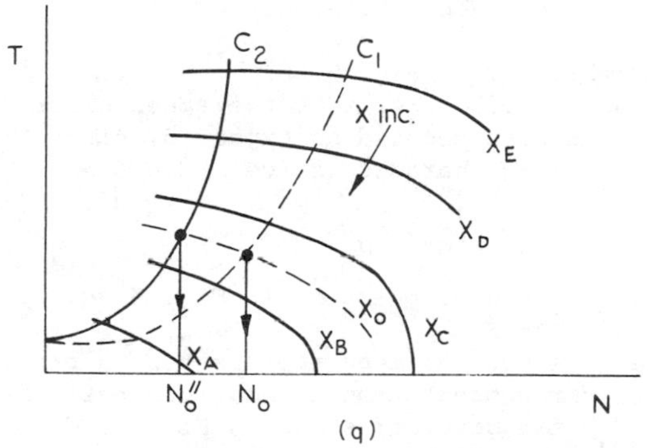

(q)

Fig. 5.2p, q. Effect of change in a characteristic

$$\Delta\beta/\Delta N = -k,$$

$\Delta\beta$ again being measured from the value β_o of valve opening at speed N_o. Thus

$$\Delta\beta = \beta - \beta_o = -k\ (N - N_o) = k\ (N_o - N),$$

in which $(N_o - N)$ is termed the speed error. This system exemplifies error - proportional control, since a correction is applied to the control valve which is *proportional* to the error in speed.

Using the method adopted in section 5.1 we may write, for a sudden change of load ΔR,

$$\Delta T_c - \Delta T_L = J\ \delta(\Delta N)/\delta t,$$

in which $\Delta T_c = C_N\ \Delta N + C_\beta\ \Delta\beta = C_N\ \Delta N + C_\beta k\ (N_o - N)$

and $$\Delta T_L = L_N \, \Delta N + L_R \, \Delta R.$$

The following equation thus applies,

$$J \, \delta N/\delta t + (L_N - C_N + kC_\beta)N = N_o \, (L_N - C_N + kC_\beta) - L_R \, (R - R_o). \quad (5.2.1)$$

This may be compared with equation (5.1.4) and, using the more concise form, we may again write

$$J \, \delta N/\delta t + bN = C.$$

It may be seen that b in equation (5.2.1) is greater than its value in equation (5.1.4) giving greater stability to the system. More importantly, the final steady speed N_o' now becomes, on putting $\delta N/\delta t$ equal to zero,

$$N_o' = \frac{b}{C} = N_o - \frac{L_R \, (R - R_o)}{L_N - C_N + kC_\beta},$$

that is the speed error,

$$N_o' - N_o = \frac{- L_R \, (R - R_o)}{L_N - C_N + kC_\beta} .$$

This is illustrated graphically in Fig. 5.2p.
Comparison with equation (5.1.7) shows that the speed error owing to a change $(R - R_o)$ in resistance has been reduced as expected, since, from an examination of the converter and load characteristics,

$$L_N - C_N > 0.$$

The latter speed error is shown graphically in Fig. 5.2q.

In this section the governor has been represented as providing proportional control. A more accurate representation of governor performance may be achieved as follows. For the governor shown in Fig. 5.2d the centrifugal force F due to each flyweight m rotating at N rad/s is given by

$$F = mN^2r,$$

where the radius of rotation r of the governor flyweights is allowed to vary and is restricted only by the resisting force in the spring k_g. Also $x = Br$ say, where B is a constant depending on the lengths of the flyweight arms.

Thus $$F = mN^2x/B.$$

Also the force F_s felt by the spring is proportional to the centrifugal force F. Thus $F_s = AF$ where A again depends on the lengths of the arms.

Hence $$F_s = AmN^2x/B.$$

Consider a small change in speed ΔN producing a small change Δx in x, and a small change ΔF_s in F_s.

Thus $$\Delta F_s = \left.\frac{\partial F_s}{\partial N}\right|_x . \Delta N + \left.\frac{\partial F_s}{\partial x}\right|_N . \Delta x.$$

Now $\Delta F_s/\Delta x$ is the spring stiffness k_g and hence

$$k_g = \left.\frac{\partial F_s}{\partial N}\right|_x \cdot \frac{\Delta N}{\Delta x} + \left.\frac{\partial F_s}{\partial x}\right|_N$$

or $$\Delta N/\Delta x = (k_g - AmN^2/B)/(2AmNx/B)$$

giving $$\Delta x/\Delta N = 2mNx/(Bk_g/A - mN^2).$$

At a steady speed N_o, let $x = x_o$.

Thus $$\left.\frac{\Delta x}{\Delta N}\right|_{N=N_o} = \frac{2mN_o x_o}{Bk_g/A - mN_o^2} = k\text{, say.}$$

This x - N relationship is only valid for small excursions of speed about the equilibrium speed N_o. Then, by integration, $x = kN + E$, where E is a further constant, which is found again applying the condition that $x = x_o$ at $N = N_o$,

that is $$E = x_o - k\,N_o,$$

whence $$x = k\,N + x_o - k\,N_o,$$

that is $$x - x_o = k\,(N - N_o),$$

giving $$\beta - \beta_o = k(N_o - N).$$

This is the equation that has been used here to describe governor performance.

Inertia effects also manifest themselves in governor performance, particularly during changes of speed. These arise owing to the mass of the flyweights and sleeve, in which damping also exists, and are usually dealt with in books on control.

5.3 Further Examples of Control Systems

In the last section we discussed the control of a prime-mover, that is a converter using a natural energy supply and providing rotational power. In that case it was an engine. Prime movers can take many forms such as steam and gas turbines, water turbines, petrol, diesel and steam engines, water-wheels and windmills. Equally control systems can be envisaged where the control device lies in the transmission between a converter and its load. For example, the overall speed-ratio of the Ward-Leonard drive of Fig. 3.1i can be controlled automatically by utilising a small d.c. generator, known as a tachogenerator as a transducer. A small subsidiary drive is then taken from the Ward-Leonard output shaft and acts as a speed-sensor. This drive provides an input to the tachogenerator, which in turn provides a voltage output to an amplifier. This amplifier responds by changing the current I_{fg} to the field-winding of the Ward-Leonard generator, thereby producing a new final output speed N_m (equation (3.1.6)). A hydrostatic speed-control system can also be constructed, again by using an output-speed sensor, this time a governor say, and employing the subsequent error signal (proportional to the difference between the actual speed and the desired speed) to actuate the swash-plate of a hydraulic pump.

For both the above systems the final output torque from the variable-speed device is a function of prime-mover speed as well as of the output speed and of the value of the controlling parameter. Thus for the Ward-Leonard speed control, from equation (3.1.6)

$$T_m = f\ (N_g,\ N_m,\ I_{fg}),$$

giving
$$\Delta T_m = \frac{\partial T_m}{\partial N_g}\bigg|\ \Delta N_g + \frac{\partial T_m}{\partial N_m}\bigg|\ \Delta N_m + \frac{\partial T_m}{\partial I_{fg}}\bigg|\ \Delta I_{fg}.$$

This assumes that the Ward-Leonard motor field current I_{fm} remains unaltered. Similarly, for the hydrostatic speed-control, from equations (3.1.2) and (3.1.4),

$$T_o = k_{15}\ P - k_{16}\ N_o$$

$$= (k_{15}/k_4)\ (k_5 N_i - Q) - k_{16} N_o.$$

Thus
$$\Delta T_o = \frac{\partial T_o}{\partial N_i}\bigg|\ \Delta N_i + \frac{\partial T_o}{\partial N_o}\bigg|\ \Delta N_o + \frac{\partial T_o}{\partial Q}\bigg|\ \Delta Q$$

in which Q depends upon the pump swashplate setting.

Comparing the Ward-Leonard and hydrostatic speed-controls, it so happens that, typically,

$$\frac{\partial T_o}{\partial N_o} \simeq 5\ \frac{\partial T_m}{\partial N_m}\ .$$

Since both are equivalent to C_N in equation (5.1.4) and both are negative, the hydrostatic drive can be seen to offer advantages over the Ward-Leonard drive in system stability.

In many applications it is important to minimize weight and here the hydraulic drive again scores, having a power-to-weight ratio typically ten times that of an electrical drive, although the maximum power capability of the latter is much greater. The reason for the power-to-weight ratio advantage of the hydraulic drive arises from the fact that the magnetic materials used in electrical systems suffer a saturation effect and a limit is reached beyond which any increase in demand can be met only by an increase in size. The comparable limit in hydraulic drives is set by the stresses which the containing materials can withstand owing to fluid pressure and here the mass effectiveness of mechanical containment scores heavily. A high power-to-weight ratio usually implies a high available acceleration of moving parts and this is especially beneficial where a rapidly-acting system is needed.

In marine and aircraft propulsion systems there is an inter-relation between propeller speed, propeller efficiency and propeller blade-pitch. Thus, when the propeller speed is reduced by reducing the fuel supply to the main-engine say, it is beneficial to alter the blade-pitch to maintain maximum propeller efficiency. This is usually carried out automatically in a so-called position-control system. Here a measuring device measures the propeller speed and this is converted to desired pitch p_1. This desired pitch is compared in a comparator with the actual pitch p (obtained by another measuring device), and the error $p - p_1$ is fed to a transducer, which alters the angle of the swashplate in a hydraulic pump. This pump delivers oil to a hydraulic motor, which in turn drives the pitch-mechanism of the propeller to alter the pitch p towards the desired value p_1.

Windmills and waterwheels are distinguished by the fact that the immediate source of power is a dynamic natural source, which, unlike a supply of fuel, cannot itself be controlled. In the case of the windmill the first requirement is to point the sails into the direction of the wind; the second is to control the effect of the wind if the speed of the sails becomes too great. The first is satisfied by the so-called fan tail which in its simplest form consists of a wind-vane mounted parallel to the axis of rotation of the main sails. The second requirement is achieved by allowing the wind to produce alterations in blade pitch to reduce efficiency and hence speed or by reducing the area of sail-frame covered by canvas.

Worked example The heat lost by radiation from a domestic hot-water heating system is proportional to the difference between the temperature of the water in the system and the air temperature in the house. When 0.29 l of oil is being burnt for which 1 l gives 33500 kJ the water temperature is 65°C when the house temperature is 15°C. The boiler efficiency is 75% and the thermal capacity of the system is equivalent to 455 l of water. The house temperature now falls quickly to 10°C.

(a) What will be the final water temperature if the boiler output remains unaltered?

(b) To effect some control a proportional controller is fitted such that the fuel supply is increased by 0.525 litres/hour for each 10°C reduction of the water temperature below 65°C.

Deduce the differential equation in terms of water temperature. How much more fuel will be burnt per hour and what will be the final water temperature?

Solution (a) Heat transferred from the boiler to the heating system

$$= 0.29 \times 33500 \times 0.75 \text{ kJ/h} = 7300 \text{ kJ/h}.$$

Under initial equilibrium conditions,

$$7300 = k_s\ (\psi - \psi_h) = k_s\ (65 - 15) = 50k_s,$$

where ψ is the water temperature, ψ_h is the house temperature and k_s is a constant of proportionality.

Hence
$$k_s = 146 \text{ KJ/h}^\circ\text{C}.$$

When the house temperature falls to 10°C, we have

$$7300 = k_s\ (\psi' - \psi_h) = 146\ (\psi' - \psi_h),$$

where ψ' is the new final water temperature.

Hence
$$\psi' = 60^\circ\text{C}.$$

(b) The heat store here is the water in the heating system and we may write

$$q_c - q_L = m\ C_H \ .\ \dot{\psi},$$

where q_c is the heat output from the converter (the boiler), q_L is the heat demand of the load, m is the mass of water in the heating system (= 455 x 1 kg) and C_H is the specific heat of the water,

$$C_H = 1 \text{ KJ/kg } ^oC.$$

Hence $$\Delta q_c - \Delta q_L = mC_H \ \delta(\Delta\psi)/\delta t \qquad (5.3.1)$$

in which $$\Delta q_c = (\partial q_c/\partial f)\ \Delta f = (\partial q_c/\partial f)\ .\ (\partial f/\partial\psi)\ \Delta\psi$$

and f denotes fuel rate.

Now $$\partial q_c/\partial f = 33500 \times 0.75 \text{ KJ/l},$$

$$-\ \partial f/\partial\psi = 0.525/10 \text{ l/h}^oC$$

and $$\Delta\psi = \psi - \psi_o = \psi - 65^oC.$$

Thus $$\Delta q_c = 1310\ (65 - \psi) \text{ kJ/h}.$$

Also $$\Delta q_L = \frac{\partial q_L}{\partial\ (\psi - \psi_h)}\ \Delta\ (\psi - \psi_h) = k_s\ \left|(\psi - \psi_h) - (\psi_o - \psi_{ho})\right|$$

$$= -\ 146\ \left|(\psi_o - \psi) - (\psi_{ho} - \psi_h)\right|$$

$$= -\ 146\ \left|(65 - \psi) - (15 - 10)\right| \text{ KJ/h}.$$

Thus the governing equation becomes, from equation (5.3.1),

$$1310\ (65 - \psi) + 146\ (65 - \psi) - 146\ (15 - 10) = mC_H\dot{\psi} = 455 \times 1 \times \dot{\psi} \text{ kJ/h}$$

or $$455\dot{\psi} + 1456\psi = (1456 \times 65) - (146 \times 5) = 95370.$$

This is the required differential equation. The final water temperature is found by putting $\dot{\psi} = 0$, whence $\psi = 64.5^oC$. The extra fuel burnt per hour has already been defined as

$$\Delta f = (\partial f/\partial\psi)\ .\ \Delta\psi = (0.525/10)\ (65 - 64.5) = 0.02625 \text{ l/h}.$$

5.4 Integral Control

Consider again the basic shortcoming of proportional control in not eliminating completely the unwanted error. For the case of the governor system discussed in section 5.2 the reason can be found in the fact that the throttle valve of Fig. 5.2d would return to its original position, if the speed of the converter returned to its desired value. How then, can we ensure that the valve does not return? Fig. 5.4a shows roughly the desired speed-time relationship upon the sudden application of a more demanding load characteristic. This illustrates an initial speed drop after which the system recovers. Fig. 5.4b shows what would happen to the governor sleeve, and the throttle valve would have to temporarily open since the sleeve and valve are rigidly linked. However, the restoration of the valve to its original position would recreate the original problem since the original fuel rate would be restored and oscillation of speed, sleeve position and valve position could ensue. If, however, we could obtain a graph similar to that of Fig. 5.4c in coordinates of valve *velocity* $\dot{\beta}$ and time the problem would be alleviated. Then the *area* under the curve would be a measure of valve opening and would remain positive after a change of load characteristic. This means that the valve would remain in its new position, thus allowing the required engine speed to be maintained. The sleeve would then have been allowed to return to its original position without

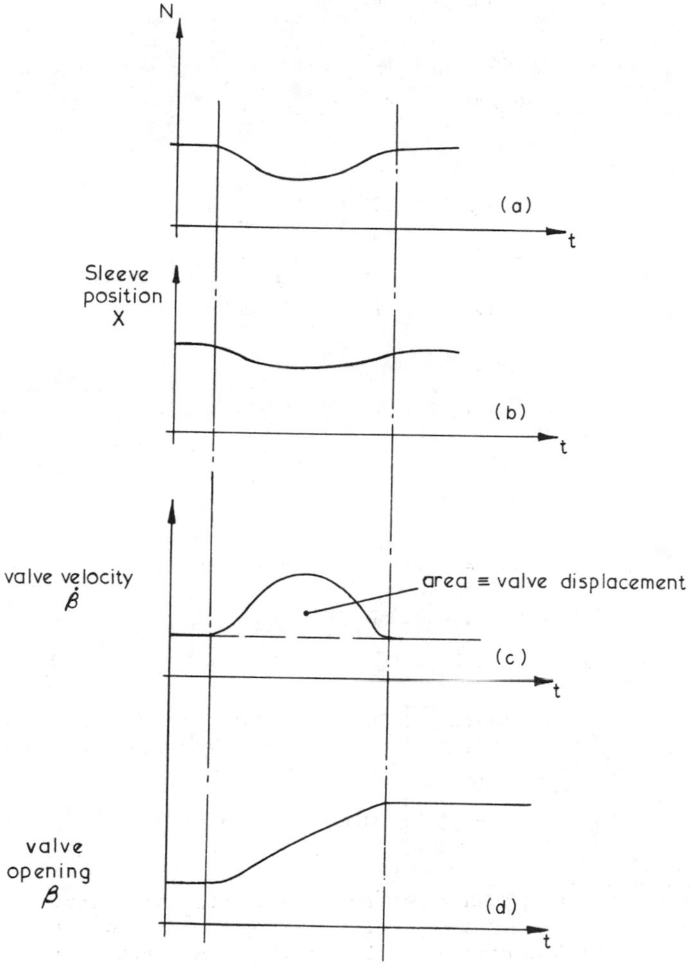

Fig. 5.4a - d. Integral control

compelling the valve to return.

The problem then is to design an intermediate device which will receive an input displacement (proportional to drop in speed) having a time variation of the form shown in Fig. 5.4b and produce an output to the control valve having velocity and displacement variations of the forms shown in Figs. 5.4c and d respectively.

Now the valve opening depicted in Fig. 5.4d is proportional to the *integral* of the input displacement of Fig. 5.4b, so we need an integrating element to perform the conversion. Figs. 5.4e, f and g show respectively fluid, mechanical and electrical versions of such an element. For the fluid version, the oil flow Q' through the spool valve into the main cylinder is given by

$$Q' = k' x',$$

where k' is known as a port constant. Also, for the main piston,

$$Q' = A \dot{\beta}.$$

Hence $$k' x' = A \dot{\beta}$$

or $$\beta = \beta_o + \frac{k'}{A} \int x'dt,$$

where β_o is the valve opening under equilibrium conditions, defined at t = 0, say.

But $$x' = - x$$

and so $$\beta = \beta_o - \frac{k'}{A} \int xdt.$$

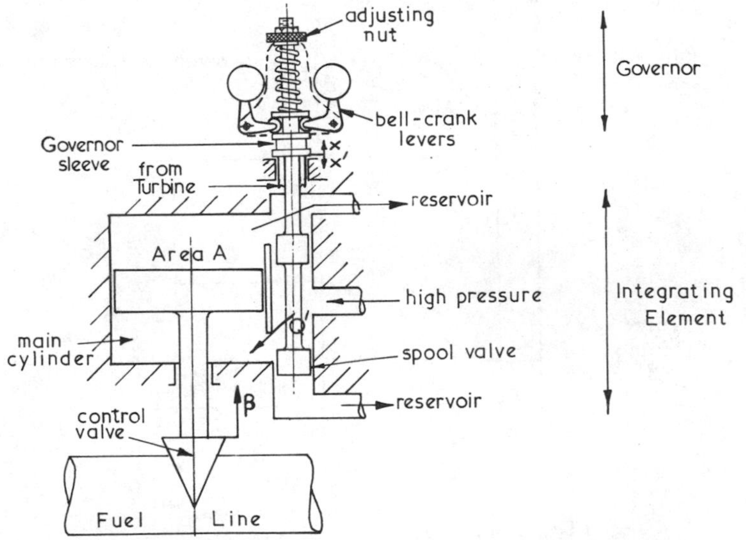

Fig. 5.4e. Governor and integrating element

In the mechanical version (Fig. 5.4f), known as a ball-and-disc integrator, let the rotation of shafts A and B in time Δt be $\Delta\theta_1$ and $\Delta\beta$ respectively. Then

$$x'\Delta\theta_1 = R\Delta\beta.$$

If shaft A is being separately driven at constant speed N_1, then $\Delta\theta_1 = N_1\Delta t$ and thus

$$\Delta\beta/\Delta t = x'N_1/R,$$

giving $$\beta = (N_1/R) \int x'\Delta t + \beta_o.$$

But using the governor of Fig. 5.4e,

$$-x' = x.$$

Hence $$\beta = \beta_o - (N_1/R) \int x\Delta t.$$

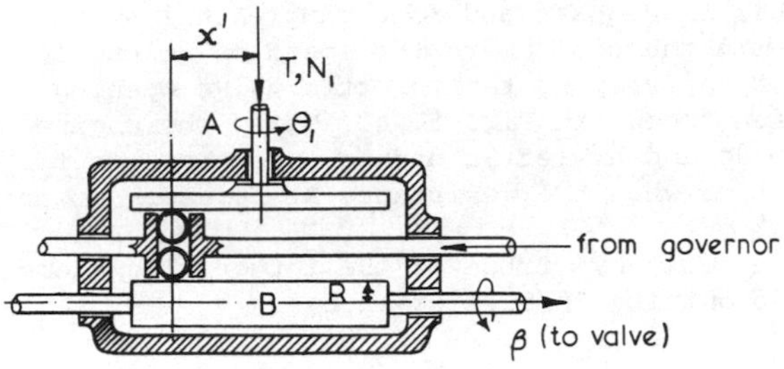

Fig. 5.4f. Ball and disc integrator

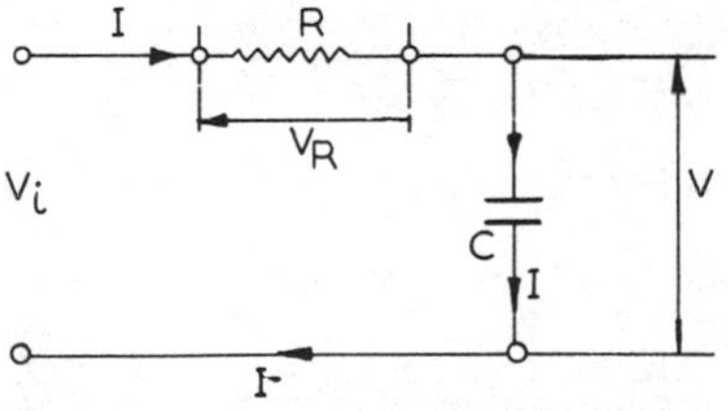

Fig. 5.4g. Integrating circuit

For the electrical version of Fig. 5.4g, known as an integrating circuit,

$$I = C\, dV/dt = V_R/R,$$

in which $$V + V_R = V_i.$$

Hence $$CR\, dV/dt = V_i - V$$

and for CR >> 1 $$V \simeq (1/CR) \int V_i\, dt + V_o.$$

If V_i and V were made proportional to x' and β respectively, such an integrating element could be used for speed control purposes. However, of these versions the fluid version is the most popular for mechanical systems, since it can be designed to embody ample amplification of force to move the control valve. Its disposition in the system would then be as shown in Fig. 5.4e.

Bearing in mind that the displacement depicted in Fig. 5.4b is proportional to speed drop and that this displacement is to be ultimately integrated, it may be worth considering an immediate integration of the speed for direct use in actuating the control valve. Such an integration could be achieved by simply measuring the number of revolutions of the output shaft in a given time and comparing this with the required number in the same time. The difference could then be used to displace the valve. Of course a source of power and an amplifier would again have to be utilised to provide the requisite force to actuate the control valve.

The introduction of an integrating element and the interest in the variation of speed and of valve opening with time means that the matching-point representation by itself is inadequate and more comprehensive techniques must be used. Consider for example the case where the speed of an engine is maintained constant at a value N_o by varying the throttle valve opening by means of the governor arrangement shown in Fig. 5.4e. Using the engine and demand characteristics of Figs. 5.2g and h, let us deduce an expression for the engine speed N as a function of time when the resistance R increases at the rate α, say.

We shall assume that, by virtue of the integrating element, the rate of change of throttle opening $\delta\beta/\delta t$ is given by

$$\frac{\delta\beta}{\delta t} = k_B \, (N_o - N).$$

We can again follow the procedure carried out in section 5.1. Let an increase in engine torque ΔT_E occur at a particular point in time.

Then
$$\Delta T_E = \left.\frac{\partial T_E}{\partial N}\right|_\beta \Delta N + \left.\frac{\partial T_E}{\partial \beta}\right|_N \Delta\beta$$

or, using the notation of section 5.1,

$$\Delta T_E = C_N \Delta N + C_\beta \Delta\beta.$$

If the increase in load torque referred to the engine is ΔT_L, then we may also write

$$\Delta T_L = \left.\frac{\partial T_L}{\partial N}\right|_R \Delta N + \left.\frac{\partial T_L}{\partial R}\right|_N \Delta R = L_N \Delta N + L_R \Delta R.$$

Now
$$\Delta T_E - \Delta T_L = J\, \delta\, (\Delta N)/\delta t$$

or
$$(C_N \Delta N + C_\beta \Delta_\beta) - (L_N \Delta N + L_R \Delta R) = J\, \delta\, (\Delta N)/\delta t,$$

where $\Delta\beta$ is measured from the valve opening β_o at the desired steady speed N_o, that is $\Delta\beta = \beta - \beta_o$. Similarly $\Delta N = N - N_o$ and $\Delta R = R - R_o$. But by virtue of the integrating element, we need to feature $\delta\beta/\delta t$ in this equation. This may be done by differentiating both sides with respect to time and we obtain

$$|C_N\, \delta N/\delta t + C_\beta\, \delta\beta/\delta t| - |L_N\, \delta N/\delta t + L_R\, \delta R/\delta t| = J\, \delta^2 N/\delta t^2.$$

It should be noticed that this equation requires the specification of a *rate* of resistance increase $\delta R/\delta t$. In this case

$$\delta R/\delta t = \alpha$$

and $$\delta\beta/\delta t = k_B\ (N_o - N).$$

Thus $$J\delta^2N/\delta t^2 + (\delta N/\delta t)\ (L_N - C_N) + C_\beta k_B N = C_\beta k_B N_o - L_R\alpha. \qquad (5.4.1)$$

The partial derivatives L_N, C_N, L_R and C_β may all be obtained from the characteristic curves of Figs. 5.2g and h, the latter two by back-plotting.

Let us now consider the solution of equation (5.4.1). From Figs. 5.2g and h it can be seen that at all values of N, and hence at $N = N_o$ (at which $\beta = \beta_o$ and $R = R_o$),

$$L_N \text{ or } \left.\frac{\partial T_L}{\partial N}\right|_R \text{ is positive,}$$

$$C_N \text{ or } \left.\frac{\partial T_E}{\partial N}\right|_\beta \text{ is negative,}$$

$$C_\beta \text{ or } \left.\frac{\partial T_E}{\partial \beta}\right|_N \text{ is positive}$$

and $$L_R \text{ or } \left.\frac{\partial T_L}{\partial R}\right|_N \text{ is positive.}$$

The final steady speed is obtained by simply writing

$$d^2N/dt^2 = 0$$

and $$dN/dt = 0,$$

whence $$N = N_o - L_R\alpha/C_\beta k_B.$$

This shows that, as α is made to approach zero, so N will approach N_o. Consider now the transient condition in acquiring this final steady state. If we first postulate that

$$L_N - C_N = 0,$$

then it is easy to see that a solution to equation (5.4.1) is of the sinusoidal form

$$N = N_o - L_R\alpha/C_\beta k_B + A \sin\ (C_\beta k_B/J)^{\frac{1}{2}}t,$$

where A is a constant depending on the value of the initial disturbance. If now we put $L_N - C_N > 0$, we know from section 5.1 that the system is stable and will return to an equilibrium condition after the disturbance. Such action manifests itself in what is known as a damped motion, either oscillatory or non-oscillatory depending upon the degree of damping provided by the difference between L_N and C_N.

Finally consider the case where the load curve remains unaltered but the governor setting is suddenly changed to ensure a higher speed N_1 of the engine, the integrator being present. Then

$$\delta\beta/\delta t = k_B\ (N_1 - N),$$

where N again has an initial condition N_o. The problem may then be solved as before to obtain the time response.

Worked example In the worked example of section 5.1 an engine and variable-pitch propeller system were considered. To this system is now added a governor and an integrating element of the form shown in Fig. 5.4e. Instead of the controller actuating a fuel valve, however, it actuates a gear which corrects the pitch of the propeller, such that the rate of change of propeller pitch dp/dt is given by $dp/dt = 7 \times 10^{-3}\ (N - N_o)$ degrees/s, where N is the engine speed in rad/s and N_o is the desired engine speed of 1140 rad/s.

Formulate the new differential equation governing the speed variation of the system when a small reduction occurs in the pitch setting of the propeller. What is the form of the resulting speed variation?

Solution The characteristics of Figs. 5.1c, d and e still apply so we may again use equation (5.1.8),

$$\Delta T_C - \Delta T_L = J\ \delta\ (\Delta N)/\delta t = 5.69\ \delta\ (\Delta N)/\delta t,$$

where
$$\Delta T_C = \left.\frac{\partial T_C}{\partial N}\right|_N \Delta N = -\ 0.8\ \Delta N$$

and
$$\Delta T_L = \left.\frac{\partial T_L}{\partial N}\right|_p \Delta N + \left.\frac{\partial T_L}{\partial p}\right|_N \Delta p$$

$$= 2.86\ \Delta N + 443\ \Delta p.$$

Hence
$$-\ 0.8\ \Delta N - 2.86\ \Delta N - 443\ \Delta p = 5.69\ \delta\ (\Delta N)/\delta t.$$

Putting
$$\Delta N = N - N_o$$

and
$$\Delta p = p - p_o$$

and differentiating both sides with respect to time gives

$$5.69\ d^2N/dt^2 + 3.66\ dN/dt + 443\ dp/dt = 0.$$

This differentiation is required to form dp/dt, which can then be related to engine speed by the constant of the integrating element. Thus

$$dp/dt = 7 \times 10^{-3}\ (N - N_o)$$

and the differential equation becomes

$$5.69\ d^2N/dt^2 + 3.66\ dN/dt + 3.101\ N = 3.101\ N_o.$$

The engine speed N has a damped oscillatory transient and a final steady value of N_o = 1140 rad/s. This contrasts with the solution obtained in section 5.1 which gave a steady value of 1164.4 rad/s. It should also be noted that the oscillation frequency is increased with increase in

$$\left.\frac{\partial T_L}{\partial p}\right|_N$$

and with increase in the constant of the integrating element.

Summarising, with error-proportional control the governing equation is, from equation (5.2.1),

$$JDN + N\ (L_n - C_N + C_\beta k) = (L_N - C_N + kC_\beta)\ N_o - L_R\ (R - R_o),$$

in which D is the differential operator d/dt, while with error-integral control the corresponding equation is, from equation (5.4.1),

$$JD^2N + DN\ .\ (L_N - C_N) + C_\beta k_B N = C_\beta\ k_B\ N_o + L_R\alpha.$$

Since C_β is positive the use of integral control instead of proportional control can be seen to result in *reduced* stability of the complete system, while fulfilling the very desirable function of nullifying the steady-state error. This reduction in stability can be appreciated when it is remembered that with proportional control the rate of decrease in throttle opening is proportional to the rate of increase in speed. On the other hand, with integral control the rate of decrease in throttle opening can only be achieved after a new increased speed has been reached. The problem of restoring an adequate degree of stability is often overcome by including both proportional and integral control elements in the feedback loop.

Worked example A liquid is fed into a tank (acting as a store) in which the level H is controlled by a float and integrating element linked to a regulating valve as shown in Fig. 5.4h. The tank has an outflow ΔQ_o limited by a set restrictor R having a resistance which allows an outflow proportional to the level H. In addition, the tank receives a disturbance in the form of another flow Q_2. The constants of the complete system are shown in Fig. 5.4h. Formulate the differential equation governing the level H.

Solution In this problem the energy store is the tank and this corresponds to the moment of inertia in the previous worked example. However, instead of storing a rate parameter (speed) it stores instead a potential parameter (pressure). As before we must first write an equation which features the inputs, the outputs and the store.

This is

$$Q_c + Q_2 - Q_o = A\ \delta H/\delta t,$$

or if we consider small changes in the variables, then

$$\Delta Q_c + \Delta Q_2 - \Delta Q_o = A\ \delta\ (\Delta H)/\delta t,$$

in which $\Delta Q_c = Q_c - Q_{co}$ etc., where the suffix 'o' denotes an equilibrium or desired condition. As far as the store is concerned the energy supplier is the regulating valve and the demand is the outlet restrictor R. We may thus write

$$\Delta Q_c = \left.\frac{\partial Q_c}{\partial \beta}\right|_H \Delta\beta = k_2\ (\beta - \beta_o)$$

and

$$\Delta Q_o = \left.\frac{\partial Q_o}{\partial H}\right|_\beta \Delta H = k_3\ (H - H_o).$$

Hence the governing differential equation becomes

$$k_2\ (\beta - \beta_o) + \Delta Q_2 - k_3\ (H - H_o) = A\ \delta\ (H - H_o)/\delta t = A\ \delta H/\delta t. \quad (5.4.2)$$

It now remains to determine β as a function of H. This can be done with reference to Fig. 5.4h in which it may be seen that, for the rigid link ABC,

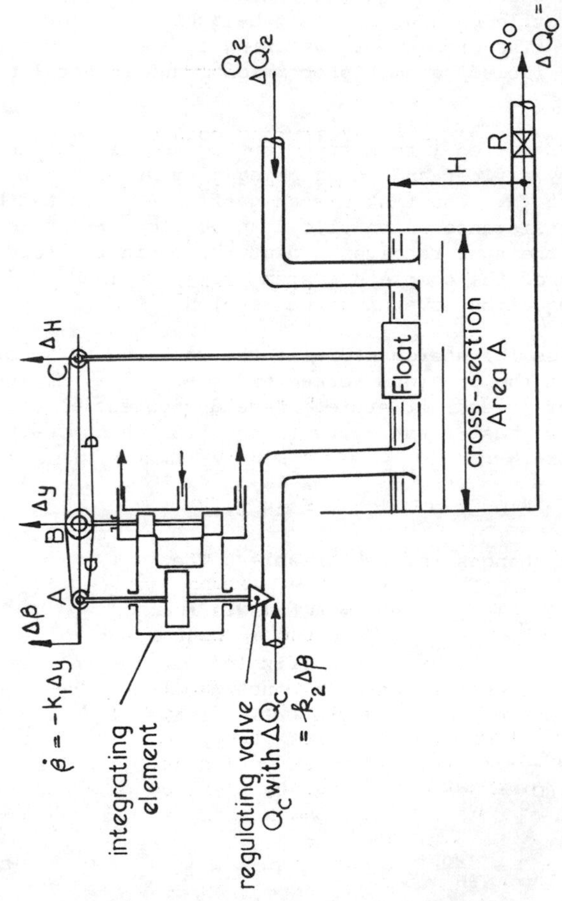

$\dot{\beta} = -K_1 \Delta y$
$\Delta\beta$
A
a
B
Δy
b
C
ΔH
integrating element
regulating valve Q_c with $\Delta Q_c = k_2 \Delta\beta$
Float
cross-section Area A
H
R
Q_2 ΔQ_2
Q_o $\Delta Q_o = K_3 \Delta H$

Fig. 5.4h.

$$\Delta y = (a\Delta H + b\Delta\beta)/(a + b),$$

in which

$$\Delta y = -\dot{\beta}/k_1.$$

Hence

$$-\dot{\beta}/k_1 = \frac{a\ (H - H_o) + b\ (\beta - \beta_o)}{(a + b)},$$

giving

$$-(\beta - \beta_o) = (H - H_o)/(\frac{a + b}{ak_1} D + \frac{b}{a}). \tag{5.4.3}$$

Substitution into the governing differential equation (5.4.2) then gives

$$-k_2\ a\ (H - H_o) + \Delta Q_2\ (\frac{a + b}{k_1} . D + b)$$

$$-k_3\ (H - H_o)\ (\frac{a + b}{k_1} D + b) = ADH\ (\frac{a + b}{k_1} D + b).$$

Now $\Delta Q_2 = Q_2 - Q_{20} = Q_2$, since under undisturbed conditions $Q_{20} = 0$.

Hence

$$A\ D^2H/k_1 + |Ab/(a + b) + k_3/k_1|\ DH + H\ (k_3b + k_2a)/(a + b)$$

$$= DQ_2/k_1 + bQ_2/(a + b) + H_o\ (k_3b + k_2a)/(a + b). \tag{5.4.4}$$

Had the link ABC not been present and the float connected directly to B then we would have written

$$\Delta y = \Delta H$$

and

$$\dot{\beta} = -k_1 . \Delta y = -k_1 . \Delta H = -k_1\ (H - H_o).$$

This is tantamount to putting b = 0 in equation (5.4.4), which would become

$$A\ D^2H/k_1 + (k_3/k_1)\ DH + k_2H = DQ_2/k_1 + k_2H_o. \tag{5.4.5}$$

Realistic values for the ratios $b/(a + b)$, $a/(a + b)$ and k_3/k_1 are about 0.5, 0.5 and 1.0 respectively and comparison of equations (5.4.4) and (5.4.5) then shows that the use of the link ABC has resulted in increased stability by increasing the degree of proportional and integral control.

Where stability is of paramount importance a beneficial type of control to employ is error-derivative control. This might be expected since error-integral control has been seen to reduce stability. Derivative control can be included, again in a feed-back path. For example consider the case of the position control system depicted in Fig. 5.4i. Here an electric motor C positions a load L. M is a measuring device and the transducer T produces K_v volts per radian error in load angular position θ.

Thus

$$\Delta T_c - \Delta T_L = J\ \delta\ (\Delta N)/\delta t = J\ \delta N/\delta t, \tag{5.4.6}$$

where J is the total effective inertia referred to the converter (motor) shaft and N is the converter speed. It will also be assumed that there is no change in load torque during any positional variation. Hence, if the motor output torque T_c is given by

$$T_c = K_T V,$$

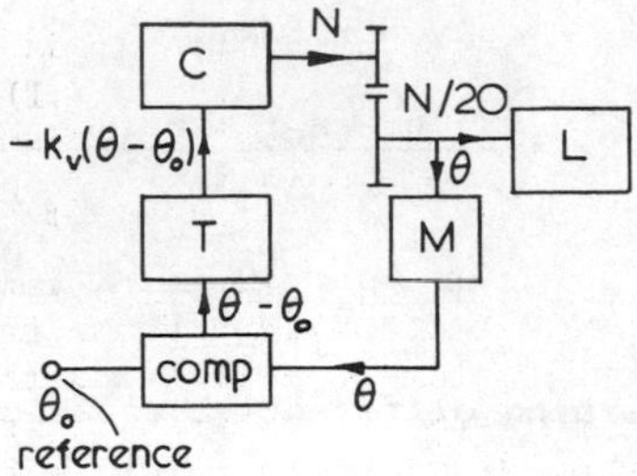

Fig. 5.4i. Position control system

then
$$\Delta T_c = \frac{\partial T_c}{\partial V} \cdot \Delta V = K_T \Delta V$$
$$= - K_T K_V (\theta - \theta_o),$$

while
$$\Delta T_L = 0.$$

Substituting into equation (5.4.6) and putting $N = 20\dot{\theta}$ gives

$$- K_T K_V (\theta - \theta_o) = 20 J d^2\theta/dt^2$$

The solution to this equation can be seen to be

$$\theta = \theta_o + A \sin t\sqrt{\frac{K_T K_V}{20 J}}$$

where A is a constant depending upon the initial condition of θ.

This solution shows that, if the reference position θ_o is altered, the angular position of the load is never steady but oscillates in a sinusoidal fashion about the desired position θ_o. Let us now introduce a further feed-back loop between the load and the converter, which provides some error-derivative control. This could be by means of a tacho-generator, which is in principle a d.c. voltage generator producing a voltage V_g proportional to its input speed N. That is

$$V_g = K_g N, \text{ say.}$$

If the input speed to the tachogenerator is arranged to be equal to the rate of change of position $\dot{\theta}$, then

$$V_g = K_g \dot{\theta}$$

and the total voltage change Δv fed back negatively into the motor C is given by

$$\Delta V = - K_v (\theta - \theta_o) - \Delta V_g$$
$$= - K_v (\theta - \theta_o) - K_g \Delta\dot{\theta}$$

$$= - K_v (\theta - \theta_o) - K_g \dot{\theta}$$

since, under initial steady conditions,

$$\dot{\theta} = \dot{\theta}_o = 0.$$

Hence $$\Delta T_C = \frac{\partial T_C}{\partial V} . \Delta V = - K_T |K_v (\theta - \theta_o) + K_g d\theta/dt|,$$

while $$\Delta T_L = 0.$$

Equation (5.4.6) becomes

$$- K_T |K_v (\theta - \theta_o) + K_g d\theta/dt| = J d^2\theta/dt^2,$$

whose solution indicates that, if θ_o is altered, the load position θ approaches θ_o in a damped oscillatory fashion. From this it is seen that derivative control has stabilised the system to advantage and will allow the load to settle to its desired position defined by the reference angle θ_o.

It is worth noting that an essential feature of a positioning system is a very high output stiffness as measured in units of torque per unit angular deflection. Now an electric motor has virtually zero output stiffness as evidenced by the ease with which its output shaft can be turned by hand with very little resistance. Hence, when used in a positioning system, some form of control as above is an absolute essential. A hydrostatic drive on the other hand has an extremely high output stiffness and can be used in an uncontrolled (open loop) configuration as in tractors, earth-moving and mining equipment.

It should be realised that many other considerations must necessarily enter into the analysis of practical control systems than have been discussed here. Such considerations include the effect of sudden large changes of load which can have important effects in an inherently non-linear system, as evidenced by the characteristics of its components. Time delays are also encountered in all mechanical engineering systems, as for example in the passage of fuel through a supply line to a diesel engine, or in the passage of air from the compressor to the combustion chamber of a gas-turbine engine. It is left to books on control to elucidate these effects.

5.5 Sinusoidal Disturbance

A great deal can usually be learnt about a system's characteristics by the application of a sinusoidal disturbance at some point. We have already observed what happens upon the application of a sudden constant disturbance. For example equation (5.1.4) was solved for a sudden change in β and in R and the result was an exponential change in speed of the system. This change was best appreciated graphically in Fig. 5.1b. In a similar way changes due to a sinusoidal disturbance can also be appreciated graphically.

Suppose we consider a system again having a governing equation of the form of equation (5.1.4), but where the valve opening β is regulated sinusoidally in accordance with the equation

$$\beta - \beta_o = \overline{\beta} \sin \omega t,$$

and R remains at its original value of R_o. Equation (5.1.4) now becomes

$$J\ \delta N/\delta t + N\ (L_N - C_N) = N_o\ (L_N - C_N) + C_\beta\ \overline{\beta}\ \sin \omega t. \qquad (5.5.1)$$

The solution of this equation consists of a so-called complementary-function and a particular integral. The former describes a transient which dies away with time and the latter a steady motion which persists. It is the latter which is of prime importance in the present context. Let us assume that this steady motion is also sinusoidal but lagging behind the disturbance by some angle, ϕ known as a phase angle. Here the motion is in terms of speed variation which may thus be written

$$N - N_o = \overline{N}\ \sin\ (\omega t - \phi).$$

Let us now try to deduce the values of $\overline{N}$ and ϕ. Substitution into equation (5.5.1) gives

$$J\ \omega\overline{N}\ \cos\ (\omega t - \phi) + \overline{N}\ \sin\ (\omega t - \phi)\ .\ (L_N - C_N) = C_\beta\ \overline{\beta}\ \sin \omega t$$

or

$$J\ \omega\overline{N}\ \left|\cos \omega t \cos \phi + \sin \omega t \sin \phi\right| + \overline{N}\ (L_N - C_N)\ \left|\sin \omega t \cos \phi - \cos \omega t \sin \phi\right|$$
$$= C_\beta\ \overline{\beta}\ \sin \omega t.$$

Equating terms in sin ωt on left and right-hand sides gives

$$J\ \omega\ \overline{N}\ \sin \phi + \overline{N}\ (L_N - C_N)\ \cos \phi = C_\beta\ \overline{\beta}. \qquad (5.5.2)$$

Equating terms in cos ωt gives

$$J\ \omega\ \overline{N}\ \cos \phi - \overline{N}\ (L_N - C_N)\ \sin \phi = 0. \qquad (5.5.3)$$

From equation (5.5.3) we obtain

$$\tan \phi = \frac{J\ \omega}{(L_N - C_N)} \qquad (5.5.4)$$

Whence
$$\sin \phi = \frac{J\ \omega}{\left|(J\omega)^2 + (L_N - C_N)^2\right|^{\frac{1}{2}}}$$

and
$$\cos \phi = \frac{L_N - C_N}{\left|(J\omega)^2 + (L_N - C_N)^2\right|^{\frac{1}{2}}}.$$

Substitution of sin ϕ and cos ϕ into equation (5.5.2) now gives

$$\overline{N} = C_\beta\overline{\beta}\ .\ \left|(J\omega)^2 + (L_N - C_N)^2\right|^{-\frac{1}{2}}. \qquad (5.5.5)$$

Equations (5.5.4) and (5.5.5) may be represented graphically in many ways but perhaps the simplest is by plotting a so-called inverse locus. To do this the value of $\overline{\beta}/\overline{N}$ is first determined. This is given by

$$\overline{\beta}/\overline{N} = C_\beta^{-1}\ \left|(J\omega)^2 + (L_N - C_N)^2\right|^{\frac{1}{2}}. \qquad (5.5.6)$$

Then, using a horizontal datum line, values of the so-called inverse response ratio $\overline{\beta}/\overline{N}$ are plotted for different values of disturbing frequency, ω using the appropriate value of ϕ as in Fig. 5.5a. By this means the inverse response locus is plotted.

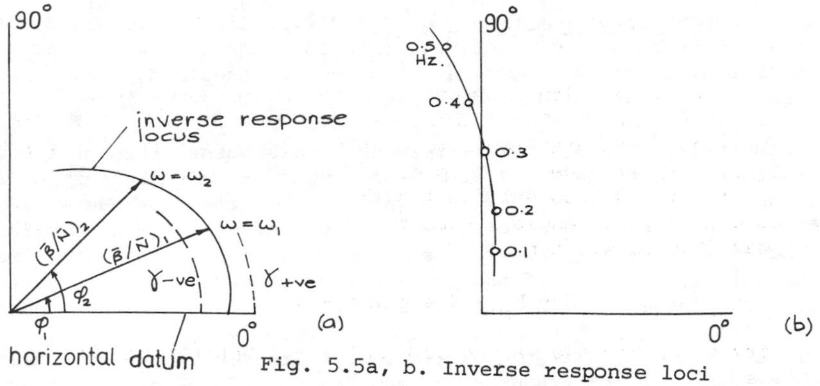

Fig. 5.5a, b. Inverse response loci

Fig. 5.5b shows an inverse response locus obtained experimentally from a heavy industrial diesel engine driving a constant-torque load on a flat portion of its torque characteristic. It is interesting to consider whether this might be of the expected form, particularly in regard to phase angle, ϕ (equation (5.5.4)). Now for a constant-torque load L_N (= $\partial T_L/\partial N$) = 0, whilst C_N (= $\partial T_C/\partial N$) is also approximately zero. Thus one would expect tan ϕ to remain at a value of approximately infinity, that is give a ϕ value of about 90°. This is seen to be roughly the case, although the locus deviates more and more from the 90° line as frequency increases. The major explanation of this deviation is that there is a delay between the movement β of the fuel-rack and the entering of the fuel into the engine. Thus strictly we cannot say that the engine torque T_C is given by

$$T_C = f\ (\beta,\ N),$$

but instead by

$$T_C = f\ (\beta,\ N,\ t_d),$$

where t_d is the delay time required for the transport of the fuel from fuel pump to engine. Thus the phase angle ϕ will be expected to be greater than 90° as is indeed the case. Since this so-called transport delay does not affect the size of the disturbance - only the timing - it can be allowed for by a small anti-clockwise rotation of the $\overline{\beta}/\overline{N}$ vector.

5.6 Quantifying the Stability

Consider again equation (5.1.4) in which we now put

$$\beta - \beta_o = \overline{\beta} e^{\gamma t} \sin \omega t,$$

that is a valve disturbance of ever-increasing excursion. Let the resulting speed variation be

$$N - N_o = \overline{N}\, e^{\gamma t} \sin\ (\omega t - \phi). \tag{5.6.1}$$

Carrying out a procedure similar to that used earlier, we obtain the following equations which are analagous to equations (5.5.4) and (5.5.6) respectively,

$$\tan \phi = \frac{J\ \omega}{J\ \gamma + (L_N - C_N)} \tag{5.6.2}$$

and

$$\overline{\beta}/\overline{N} = C_\beta^{-1}\ |(J\omega)^2 + (J\ \gamma + L_N - C_N)^2|^{\frac{1}{2}}. \tag{5.6.3}$$

The inverse response loci now depend on γ and will lie, in relation to the locus for $\gamma = 0$, as shown in Fig. 5.5a. This is evident from the fact that, if γ is positive, tan ϕ is reduced and $\bar{\beta}/\bar{N}$ is increased and, if γ is negative, tan ϕ is increased and $\bar{\beta}/\bar{N}$ reduced.

If at some value of ω an inverse response locus passes through the origin O, this means that a response, $\bar{N}$ exists for no input, $\bar{\beta}$. If the particular inverse response locus is one for which γ is negative, then the response will die away with time and the system is stable. γ is then a measure of the damping in the system. However, if γ is positive the response will increase with time, denoting instability. Hence, if the locus corresponding to $\gamma = 0$ passes above the origin, the system must be stable, whilst if it passes below, then the system is unstable. Furthermore, for a $\gamma = 0$ locus which *just* passes through the origin the value of ω at the origin is the frequency of unstable motion which the system will acquire upon the application of a small arbitrary disturbance.

5.7 Transient Performance

Let us consider again equation (5.6.1). Thus

$$\partial(N - N_O)/\partial\omega = \bar{N}\, e^{\gamma t}\, t\cos(\omega t - \phi) \tag{5.7.1}$$

and

$$\partial(N - N_O)/\partial\gamma = \bar{N}\, e^{\gamma t}\, t\sin(\omega t - \phi). \tag{5.7.2}$$

Thus changes of the output parameter $(N - N_O)$ with respect to frequency ω are of the same *magnitude* as changes of $(N - N_O)$ with respect to γ. The spacing of the loci for ω = constant and for γ = constant must therefore be equal around any point on the locus. Further the cosine and sine terms in equations (5.7.1) and (5.7.2) tell us that these spacings are at right angles. Thus once the spacings on a $\gamma = 0$ locus for changes in ω are known (Fig. 5.5a), the loci for other γ values may be sketched in as shown. Then the γ locus which passes through the pole O gives the value of γ for the system since for the locus to pass through the pole the input disturbance $\bar{\beta}$ must be zero, and so γ determines the increase or decay of any transient which the system exhibits due to some small perturbation.

5.8 Closed Loop Response

Let us now consider the application of a sinusoidal disturbance to a system embodying a feed-back path, for example the system of Fig. 5.2c. This is known as a closed-loop system to distinguish it from the open system of section 5.1. To apply the disturbance we must first break the system at some suitable point. The disturbance can then be injected at one end of the break and the response is measured at the other end of the break. Let us break the system at a point between the transducer and the control device of Fig. 5.2c, say, and inject a constant-amplitude sinusoidal disturbing force F to the control device. This will cause a sinusoidal force to emanate from the transducer, which we can call G, say. The inverse response ratio is thus F/G and since we have opened the loop to find its value, we call it the open-loop inverse response ratio, I_{OL} say.

Thus

$$I_{OL} = F/G. \tag{5.8.1}$$

Let us now close the loop and return to the system of Fig. 5.2c, complete with feedback, applying the same disturbing force F to the control device which, together with a sinusoidal force H from the transducer, produces a net force

(F + H) on the control device. The inverse response ratio is now F/H and is called the closed-loop inverse response ratio, I_{CL}, say. But (F + H) now replaces F in equation (5.8.1) while H replaces G.

Thus $$I_{CL} = F/H = (F + H)/H - 1 = I_{OL} - 1.$$

The simple nature of the relationship between I_{CL} and I_{OL} means that the locus of I_{CL} can be found very easily once the locus of I_{OL} is known (Fig. 5.8a).

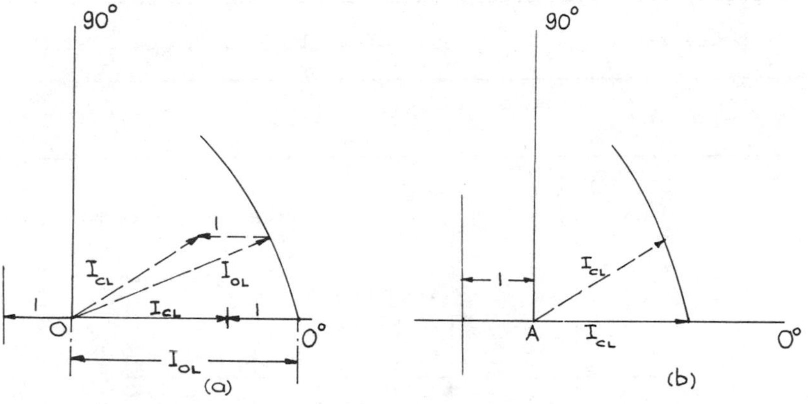

Fig. 5.8a, b. Derivation of closed-loop inverse response locus from the open-loop inverse response locus

For example, let us consider the case where $\phi_{OL} = 0$. The I_{CL} is found as shown in Fig. 5.8a by simply subtracting unity from I_{OL}, which is equivalent to moving the origin of the locus from O to A (Fig. 5.8b). By this means the I_{OL} locus will serve to describe I_{CL} without the need to redraw it. The same methods as used previously then apply for assessing stability and finding the damping and frequency of the closed-loop system.

<u>Worked example</u> An open loop frequency response test was carried out on a controlled chemical process by opening the loop at the controller output and subjecting the plant control valve to a sinusoidal air pressure of 350 N/mm^2 amplitude and at various frequencies. The air pressure output from the controller was measured in magnitude and phase relative to the input and the results are tabulated on page 150.

Draw the closed-loop inverse response locus and estimate the damping and frequency of oscillation of the closed-loop system.

frequency (c/min)	1	3	5	7	9	11
output air pressure (N/mm^2)	440	340	274	210	161	112
phase lag (degrees)	313	328	339	351	363	375

Solution The above table may be rewritten as follows:

frequency (rad/s)	0.105	0.315	0.525	0.735	0.945	1.155
inverse response ratio	0.8	1.03	1.28	1.66	2.18	3.11
phase lag (degrees)	313	328	339	351	363	375

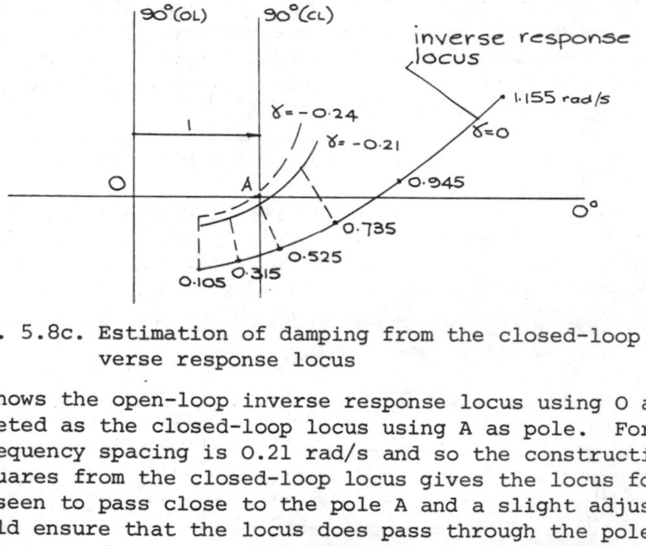

Fig. 5.8c. Estimation of damping from the closed-loop inverse response locus

Fig. 5.8c shows the open-loop inverse response locus using O as pole. This may be interpreted as the closed-loop locus using A as pole. For both loci $\gamma = 0$. The frequency spacing is 0.21 rad/s and so the construction of rough curvilinear squares from the closed-loop locus gives the locus for $\gamma = -0.21$. This locus is seen to pass close to the pole A and a slight adjustment to $\gamma = -0.24$ would ensure that the locus does pass through the pole. The corresponding frequency of damped oscillation is about 0.525 rad/s.

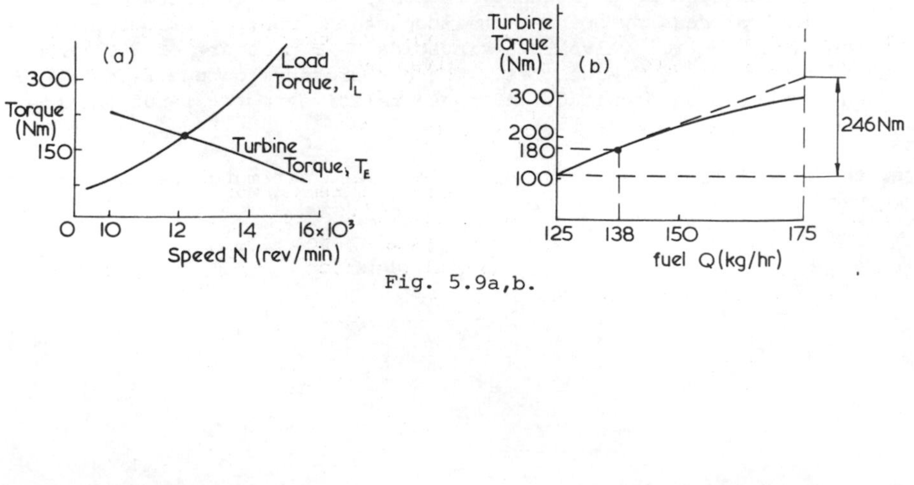

Fig. 5.9a,b.

5.9 Examples

1. A petrol engine drives a pump through a step down gearbox of 2:1. At the operating point of the pump $\partial T_p/\partial N_p = f$ and $\partial T_p/\partial P_p = k$.

The slope of the engine $T_E - N_E$ characteristic at its normal operating speed N_{Eo} is $\partial T_E/\partial N_E = -c$, and the total moment of inertia of the engine and pump referred to the engine is J. Formulate the differential equation governing the engine speed N_E for a small change ΔP_p in pump delivery pressure.

(Ans. $JDN_E + (f/4 + c)\, N_E = (k/2)\, \Delta P_p + (f/4 + c)\, N_{Eo}$).

2. A gas turbine drives a propeller. In order to control the speed of the gas-turbine a governor is used to regulate the fuel supply, the pitch-angle of the propeller being kept constant. Fig. 5.9a shows the torque speed output and demand characteristics relating to the turbine, indicating a steady turbine speed of 12000 (rev/min) when the fuel supply is 138 (kg/h). Fig. 5.9b shows the way in which the turbine torque varies with fuel supply for a turbine speed of 12000 (rev/min). The governor regulates the fuel supply such that

$$\Delta Q = - \frac{\Delta N}{12.5D + 38}\ \text{kg/h},$$

where ΔQ and ΔN are the increases in fuel rate and turbine speed in rev/min respectively. The total moment of inertia referred to the turbine shaft is 2 kg m^2. Draw a block diagram of the complete system and formulate the differential equation governing the turbine speed variation with time for a small disturbance in speed from 12000 rev/min.

(Ans. $0.07\, D^2N + 0.233\, DN + 0.2\, N = 2400$, where N is in rev/min).

3. Fig. 5.4i shows a block diagram of a position control system in which an electric motor C turns a load L to an angular position. M is a measuring device and the transducer T produces 10 volts per radian error in load angular position. The motor inertia is 1.35×10^{-5} kg m^2 and the load inertia is 1.35×10^{-3} kg m^2. The torque-speed equation for the motor is

$$T_c = 1.356 \times 10^{-3}\ (V - 0.1\ N)\ \text{N m}$$

where V is the input voltage and N is the motor speed in rad/s.
Formulate the differential equation governing the angular position θ of the load for a 0.1 rad step change in reference position θ_o. Assume that there is no change in load torque during the motion. What can be concluded about the rôle of droop in the T_c - N characteristic of the motor?

(Ans. $0.25\, D^2\theta + 2\, D\theta + 10\theta = 1$).

4. The gas furnace of a heating installation of thermal capacity C supplies heat to the radiator water at a rate q given by

$$q = \frac{K}{TD + 1} \cdot (\theta_o - \theta),$$

in which K and T are constants, θ is the actual temperature of the water in the radiators and θ_o is the temperature set on the regulator. The gas supply cuts out if θ exceeds θ_o. The radiator system rejects heat at a rate $K_1\theta$ where K_1 is a constant. Deduce the differential equation expressing θ in terms of θ_o when the furnace is operating.

(Ans. $TD^2\theta + D\theta\ (C + K_1T) + \theta\ (K + K_1) = K\theta_o$).

5. A water-wheel has a torque-speed output characteristic which passes through the following points:

Speed N_w rev/s	0	1	2	3	4	5
Torque T_w Nm	140	140	135	115	70	0

(a) It is desired to utilise the maximum power output. At what speed should it be run?

(b) The water-wheel drives a mill whose torque-speed demand characteristic is given by

$$T_m = 70\ N_m^{\ 2}\ \text{Nm},$$

where N_m is the mill speed in rev/s. Calculate the gear-ratio, G (= N_m/N_w) required to obtain maximum power at the mill, assuming no transmission losses.

(c) Formulate and solve a differential equation using the gear-ratio deduced above to describe the variation in mill-speed with time, when a *small* increase of 0.5 m/s now occurs in the speed of the water entering the wheel. The relation between wheel output-torque T_w and water-speed v may be assumed to be

$$T_W = 23\ v\ \text{Nm},$$

where v is in m/s. The effective moment of inertia at the mill-shaft may be taken to be 0.375 kg m^2 and at the water-wheel shaft 0.1 kg m^2.

(Ans. (a) 3 rev/s, (b) 0.567 (c) $2\pi\dot{N}_m + 515\ N_m = 900$).

6. A steam engine drives a generator through a speed increasing gearbox of ratio G. The generator delivers power to a load resistor R. The torque-speed output characteristic of the engine is given by

$$T + eN = f\beta$$

where β is the throttle setting.

The torque-speed demand characteristic of the generator is given by

$$aN - bT = V$$

where V is the output voltage.

The voltage-current output characteristic of the generator is given by

$$V + CI = aN$$

where N is the generator speed.

Show that if small increases $\Delta\beta$ and ΔR occur simultaneously then the increase in engine speed, ΔN is given by

$$\Delta N = \frac{f \cdot \Delta\beta + \dfrac{G^2\ aN/cb}{(1 + R/c)^2} \cdot \Delta R}{e + \dfrac{G^2\ a/b}{(1 + R/c)}} .$$

If proportional control is applied in the form of a governor such that $\Delta\beta = - K\ \Delta N$ show that the engine speed error **E** is given by

$$E = \frac{G^2\ aN/cb}{(1 + R/c)^2 \left[e + \dfrac{G^2 a}{b\ (1 + R/c)} + fK \right]} \cdot \Delta R.$$

7. A small steam turbine drives a generator through a gearbox. The generator delivers power to an electrical load which may be regarded as purely resistive of value 3 ohms.

The torque-speed output characteristic of the engine may be approximated by

$$T + 0.55 \times 10^{-3}\ N = 0.36\ \beta$$

in the operating range.

The torque-speed demand characteristic of the generator is given by

$$0.14\ N - 0.2\ T = V$$

and the voltage-current output characteristic of the generator by

$$V + 0.25\ I = 0.14\ N.$$

In all the above equations, T represents torque in Nm, N speed in rev/min, β throttle opening in degrees, V voltage in volts and I current in amps.

(a) Deduce the gear ratio output speed/input speed such that the steam turbine will deliver maximum power to its load when the throttle setting is 10 degrees.

(b) When running at this maximum power point small increases of 1 degree and 0.1 ohm now occur simultaneously in the throttle-setting and resistive load respectively. Deduce the resulting change in turbine speed, assuming the above gear ratio.

(Ans. 0.105; 367 rev/min increase).

8. A turbine drives a variable-pitch propeller at 990 rev/min through a speed-reducing gearbox of ratio 11:1. The turbine torque-speed output characteristic at a constant throttle opening is shown in Fig. 5.1c. The propeller torque-speed demand characteristic for the particular pitch setting used is shown in Fig. 5.1d. The moments of inertia of the turbine rotating parts and of the propeller are 4.6 and 130 kg m^2 respectively.

When a small change is made to the pitch setting of the propeller and the throttle opening remains constant the propeller speed increases to 1012 rev/min.

Proportional control is now applied by means of the governor shown in Fig. 5.2d in which the link arms are equal. The steady-state error in propeller speed is 2 rev/min when the same change in pitch setting is again made.

The same governor is now connected to the control valve via an integrating element (Fig. 5.4e) such that $\dot{\beta} = -0.0164\ X$.

Deduce the differential equation describing the performance of the complete system with integral control. Work from first principles and assume linear performance throughout.

(Ans. $5.67\ d^2N/dt^2 + 23.34\ dN/dt + 3.82\ N = 63$ with N in rev/s).

9. A process control loop was broken at a suitable point and an input $X_A \sin \omega t$ applied at one end of the break. From the other end an output $X_B \sin (\omega t - \phi)$ was obtained in accordance with the following table:

ω rad/s	0.2	0.4	0.6	0.8	1.0	1.2
X_A/X_B	0.2	0.46	0.8	1.3	2.0	2.94
ϕ°	290	313	332	347	360	370

Draw the inverse response locus for the system in closed-loop and deduce approximately the damping and frequency of oscillation.

(Ans. $1.22s^{-1}$, $0.74s^{-1}$.)

INDEX

PERGAMON INTERNATIONAL LIBRARY
of Science, Technology, Engineering and Social Studies
The 1000-volume original paperback library in aid of education, industrial training and the enjoyment of leisure
Publisher: Robert Maxwell, M.C.

The Characteristics of Mechanical Engineering Systems

THE PERGAMON TEXTBOOK
INSPECTION COPY SERVICE

An inspection copy of any book published in the Pergamon International Library will gladly be sent to academic staff without obligation for their consideration for course adoption or recommendation. Copies may be retained for a period of 60 days from receipt and returned if not suitable. When a particular title is adopted or recommended for adoption for class use and the recommendation results in a sale of 12 or more copies, the inspection copy may be retained with our compliments. The Publishers will be pleased to receive suggestions for revised editions and new titles to be published in this important International Library.

Other Titles of Interest in the Pergamon International Library

BENSON	Advanced Engineering Thermodynamics, 2nd Edition
BRADSHAW	Experimental Fluid Mechanics, 2nd Edition
BRADSHAW	An Introduction to Turbulence and its Measurement
BUCKINGHAM	The Laws and Applications of Thermodynamics
DANESHYAR	One-Dimensional Compressible Flow
DIXON	Fluid Mechanics, Thermodynamics of Turbomachinery, 2nd Edition
DIXON	Worked Examples in Turbomachinery (Fluid Mechanics and Thermodynamics)
HAYWOOD	Analysis of Engineering Cycles, 2nd Edition
MORRILL	An Introduction to Equilibrium Thermodynamics
PEERLESS	Basic Fluid Mechanics
SCHURING	Scale Models in Engineering
THOMA	Bond Graphs

The Characteristics of Mechanical Engineering Systems

R. HOLMES

School of Engineering & Applied Sciences
University of Sussex

PERGAMON PRESS

OXFORD · NEW YORK · TORONTO · SYDNEY · PARIS · FRANKFURT

UNIVERSITY OF
STRATHCLYDE LIBRARIES

U.K.	Pergamon Press Ltd., Headington Hill Hall, Oxford OX3 0BW, England
U.S.A.	Pergamon Press Inc., Maxwell House, Fairview Park, Elmsford, New York 10523, U.S.A.
CANADA	Pergamon of Canada Ltd., 75 The East Mall, Toronto, Ontario, Canada
AUSTRALIA	Pergamon Press (Aust.) Pty. Ltd., 19a Boundary Street, Rushcutters Bay, N.S.W. 2011, Australia
FRANCE	Pergamon Press SARL, 24 rue des Ecoles, 75240 Paris, Cedex 05, France
WEST GERMANY	Pergamon Press GmbH, 6242 Kronberg-Taunus, Pferdstrasse 1, Frankfurt-am-Main, West Germany

Copyright © 1977 R. Holmes

All Rights Reserved. No part of this publication may be reproduced, stored in a retrieval system or transmitted in any form or by any means: electronic, electrostatic, magnetic tape, mechanical, photocopying, recording or otherwise, without permission in writing from the publishers

First edition 1977

Library of Congress Cataloging in Publication Data

Holmes, Roy, MSc.
The characteristics of mechanical engineering systems.

(Pergamon international library of science, technology, engineering, and social studies)
1. Machinery, Dynamics of. 2. Power (Mechanics)
I. Title.
TJ173.H64 1977 621 76-56248

ISBN 0-08-021033-3 (Hard cover)
ISBN 0-08-021032-5 (Flexi cover)

In order to make this volume available as economically and rapidly as possible the author's typescript has been reproduced in its original form. This method unfortunately has its typographical limitations but it is hoped that they in no way distract the reader.

FLOOD DAMAGE

2 3 JAN 1996

Printed in Great Britain by A. Wheaton & Co., Exeter

D
621.811
HOL

CONTENTS